30

GREAT MYSTERIES OF THE AIR

By the same author in Pan Books

STRIKE HARD, STRIKE SURE

CONDITIONS OF SALE

This book shall not, by way of trade or otherwise, be lent, re-sold, hired out or otherwise circulated without the publisher's prior consent in any form of binding or cover other than that in which it is published and without a similar condition including this condition being imposed on the subsequent purchaser. This book is published at a net price, and is supplied subject to the Publishers Association Standard Conditions of Sale registered under the Restrictive Trade Practices Act, 1956.

RALPH BARKER

GREAT MYSTERIES OF THE AIR

UNABRIDGED

PAN BOOKS LTD: LONDON

First published 1966 by Chatto & Windus Ltd.

This edition published 1968 by Pan Books Ltd.,
33 Tothill Street, London, S.W.1

ISBN 0 330 02096 X

2nd Printing 1972

Maps by J. R. Flower © Chatto and Windus Ltd., 1966
Text © Ralph Barker, 1966

PRINTED AND BOUND IN ENGLAND BY
HAZELL WATSON AND VINEY LTD
AYLESBURY, BUCKS

ACKNOWLEDGEMENTS

A great many people have helped me in gathering material for these stories, and I have gone to numerous sources for information and assistance. Among individuals and individual organizations I should particularly like to mention the following:

Air Commodore Freddie West, VC, and P. R. Reid, the author; Air Commodore Sir Vernon Brown; Sir Roy Fedden; 'Toby' Charlton; and Harwood Steele, author and journalist, all of whom gave me valuable help in compiling the story on Colonel Minchin, Leslie Hamilton and Princess Loewenstein-Wertheim.

R. F. Little and Captain R. H. McIntosh, for their personal accounts of the details as known to them of the last flight of Captain Alfred Loewenstein.

British Overseas Airways Corporation, for permission to see files and papers concerning the loss of the Imperial Airways Argosy *City of Liverpool* and the wartime KLM Dakota G-AGBB.

Beaverbrook Newspapers, for permission to use Bill Lancaster's diary as source material; also the French Embassy in London, for providing information on the known details of Bill Lancaster's last flight.

Flight Lieutenant Andrew Jack, for permission to draw on his account of the last flight of the Duke of Kent; also Group Captain N. M. Maynard, for background detail.

The Information Department of the British Petroleum Company, for help with details of the finding of the B24 Liberator bomber *Lady Be Good*; also Alfred Respinger, for his maps and description of the terrain.

T. Scott-Chard, of BOAC, for his account of events at Lisbon before Leslie Howard's last flight; and Guy R. Holmes, for his account of how his brother, the late Reverend Arthur Holmes, came to miss the flight.

Cecil Madden, Maurice Gorham and Charles Max-Muller

for their accounts of the work of Glenn Miller with the BBC; Joyce Rowe, BBC Sound Publicity Officer; John Quenby and his family, of Twin Woods Farm, Bedford; Denis J. Corley; and the Department of the Air Force, Washington.

Captain Frank Griffin, of BOAC, for his account of his parallel flight in a Lancastrian on the day the British South American Airways Tudor *Star Tiger* disappeared.

Captain Ian Harvey, of British European Airways, and Sue Cramsie, formerly of British European Airways, for their accounts of the Viking flight from Northolt of 13th April 1950.

I should also like to thank the Librarian and staff of various libraries, especially the Royal Aeronautical Society, the Central Library of the Ministry of Aviation, the Air Ministry Library, the United States Information Service Library, and the Library of Beaverbrook Newspapers. In addition I must mention my great indebtedness to the Library of the British Museum, and the Newspaper Library at Colindale. I have also made use of microfilm of the *New York Times* kept at Printing House Square.

I have received valuable help from the Air Historical Branch and Public Relations Branch of the Ministry of Defence (Air), and from the Public Relations staffs of BOAC and BEA.

Civil Aircraft Accident Reports of the then Ministry of Civil Aviation have been an invaluable source for three of the stories – 'The Long Silence of *Star Tiger*', 'A Mistake on a Map', and 'The Infernal Machine'.

Among the many books consulted for factual background I must particularly mention *That's My Story*, by Douglas Corrigan, published in 1939 by Robert Hale, from which many of the facts about Corrigan and his Atlantic flight are taken.

Copyright of photographs is acknowledged as follows: 11, *Associated Press*. 1, 2, and 6, *United States Air Force*. 5, *Press Association*. 7, 9, and 12, *London Express News and Feature Services*. 8 and 10, *BEA*.

Copyright of map on p. 119, *HMSO*. R. B.

CONTENTS

ILLUSTRATIONS IN PHOTOGRAVURE

MAPS

I

AMELIA EARHART

ON either side of the narrow strip of runway pressed the dense, impenetrable vegetation of the New Guinea jungle. At the bottom of that runway, the silver-fuselaged, twin-engined Lockheed Elektra paused for a moment before turning into wind. At the opposite end, a thousand yards distant, forest and runway ended abruptly where the land fell away sheer to the sea. Beyond the cliff edge lay the Pacific, 7,000 miles of it, separating the crew of the Lockheed from Oakland, California – and home.

A month earlier, on 1st June 1937, the crew of the Lockheed, pilot and navigator, had started out from that same Oakland on a round-the-world flight, turning their backs on the Pacific and heading east across the United States to Miami. Then their route, hugging the Equator wherever possible, had taken them south-east to Brazil, over the South Atlantic, across the waist of Africa, and thence via Karachi, Rangoon, Singapore, Sourabaya and Port Darwin to New Guinea. Twenty-two thousand miles lay behind them. Ahead lay the vast Pacific.

The navigator of the Lockheed was one of the most skilled and experienced in America, Captain Fred Noonan. The pilot was Amelia Earhart.

Amelia Earhart had first won fame in 1928 when, as a passenger in a three-engined Fokker, she had become the first woman to fly the Atlantic. Something about that triumph had left her unsatisfied. It was not until 1932, when she became the first woman to fly the Atlantic solo, breaking the speed record as she did so, that she felt she had earned her fame. A woman of sensitivity and intelligence, she had married G. P. Putnam, the New York publisher, three years after her first Atlantic flight. She was then 33. Trousered and with cropped hair, she was a boyish figure who somehow, like Katherine Hepburn, never lost her femininity. Yet she was no playgirl of the air. She spoke five languages, was a university graduate, and spent her life – when she wasn't flying – doing welfare work in the city of Boston. The pitfalls of fame and publicity did not trap her; there were never any unsavoury rumours or stories about Amelia Earhart. Without complexes or obsessions, and without any conscious aim of competing with men, she simply set out to show that women, too, could do these things.

The round-the-world flight was to be her last; she announced this before leaving Oakland. She was 39, and she felt the time had come to face middle-age. She even talked about growing old gracefully. 'But I feel,' she said, 'that I've got one more big flight in me.' So she pressed forward with her plans for a round-the-world flight. It would be something to savour and look back on in the coming years. But more important, she intensely wanted to do it. She had refused to be put off by difficulties. Indeed this was her second attempt. She had begun her first try in the reverse direction and had successfully negotiated the long ocean hop from Oakland to Honolulu. Then, leaving Honolulu for the tiny speck of Howland Island in the central Pacific, she had crashed on take-off. Now, on Friday, 2nd July 1937, she was about to take off from Lae, New Guinea, and the wheel had come full circle; once again she was taking off for elusive Howland Island. She hoped for better luck this time.

Throughout the trip she had known that this hop would be the hardest. It involved an ocean crossing of 2,550 miles, the longest ever attempted. And Howland Island, half a degree north of the Equator, one and a half miles long and half a mile wide, and only a few feet above sea-level, was little more than a sandbar in a vast expanse of ocean. The runway there had only just been built. No plane had ever landed there. Even with the aid of an expert navigator it was a fantastically difficult target to find. The flight could not have been contemplated if it hadn't been possible to arrange for an American coastguard cutter to anchor off the island and give radio assistance.

The Lockheed was heavily loaded with petrol. To make the take-off easier, Noonan and Miss Earhart had discarded all unnecessary baggage. Noonan had nothing more than a small tin box. Miss Earhart carried only a brief-case, in which she packed her papers, some extra clothing, and a toothbrush. A score of Europeans and nearly a thousand Papuans watched the silver plane glinting down the runway. Right to the end of that 1,000-yard avenue the undercarriage wheels were still turning. Then, just as the plane seemed about to sink over the edge of the cliff and disappear, it flew off gently and smoothly across the gulf. There was no circling of the town of Lae in salute; fuel was too critical for that. A small adjustment of heading soon after that easterly take-off and the plane was on course.

It was ten AM, New Guinea time. The Lockheed's cruising speed would be not less than 150 miles per hour. In conditions of still air the flight would take 17 hours. But a headwind was forecast. Captain Noonan estimated that the flight would take about 20 hours. They had fuel for 24.

The Elektra was well-equipped for long-distance flying. Noonan had all the instruments he needed for celestial navigation. In his chart-room aft of the tiny cockpit there was a Bendix radio-compass, and a powerful radio was carried for voice communication. Its range depended on atmospheric conditions and was apt to be capricious, but over water it normally gave a working range of several hundred miles.

The flight plan was that they would fly all that day by dead reckoning, checking their position if possible as they overflew the Solomon Islands. Then, through the night, Noonan would take frequent star shots. If there was cloud they would climb through it. Tomorrow morning, as they approached Howland Island, they would come within voice range of the *Itasca*, the coastguard cutter that had been positioned specially for them. The *Itasca* was equipped with a high-frequency radio direction finder. If they had any difficulty in locating the island, the *Itasca* would guide them in.

The threat of strong headwinds had delayed them at New

Guinea for 24 hours. Now, as they headed east-north-east, the weather was calm and the headwind moderate. Thirty minutes after take-off Amelia Earhart broadcast her first routine half-hourly report of progress, and for the next eight hours, while the Elektra covered over a thousand miles, these reports were picked up at Lae. They showed that four hours after take-off the weather had deteriorated. The headwinds, too, seemed to

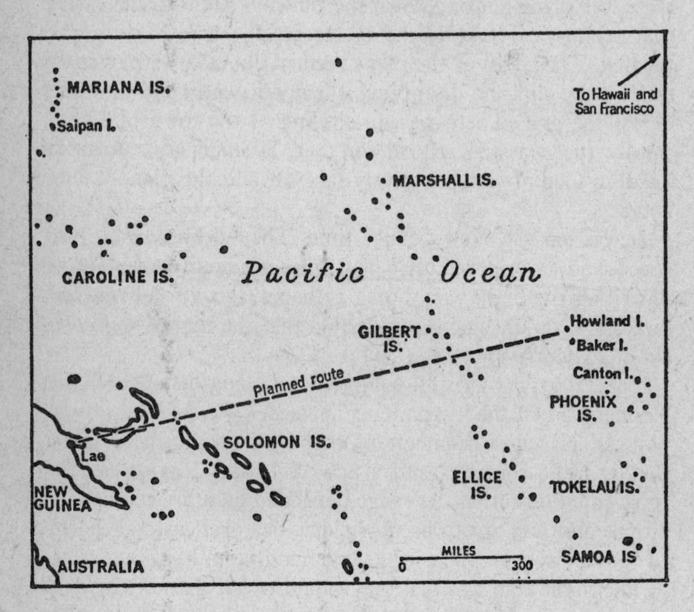

1. New Guinea to Howland Island

have stiffened. But Amelia Earhart and her companion were sitting it out at 8,000 feet, relaxed and confident. The twin-Wasp engines of the Lockheed were behaving perfectly. All the indications were that this huge stretch of water was about to be safely traversed by air for the first time.

The ground operators at Lae believed that they kept track of the Lockheed for approximately 1,200 miles. They could not fix her position at this distance, but so far as they could judge from the messages they received she was holding her course. All the time the signals were fading, and finally they disappeared. But there was nothing significant in this. The surprising thing was that the Lockheed's signals had carried so well.

Amelia Earhart was still making her half-hourly reports of progress, but here, in the middle of this long tight-rope of flight, radio communications sagged. She had lost touch with Lae, and it might be many hours before she made contact with the *Itasca*. It was now that Fred Noonan's skill as a navigator would be tested. As night enveloped the cockpit, he got busy with his sextant. On the ground at Lae he had reset his chronometers. Any undisclosed faults of timing would have rendered his reading inaccurate.

Throughout the night the cutter *Itasca* endeavoured to make contact with the Elektra. Throughout the night Amelia Earhart made her half-hourly routine calls, hoping they might be picked up. But static interference blotted out her signals. Towards morning the operators on the *Itasca* heard her voice several times, but the words were unintelligible.

The Lockheed was due to reach Howland Island at eight o'clock that morning, local time. Noonan believed that in spite of adverse winds they were ahead of schedule. At 6.15 Howland time Amelia Earhart's voice at last penetrated the crackle of the static.

'We are about a hundred miles out. Please take a bearing on us and report in half an hour. I will transmit into the microphone.'

Then the static closed in again. The operators in the *Itasca*'s radio room, excited at this contact, swung their aerials expertly, but they had no signal to work on. It was impossible to get the bearing of the Elektra.

For the next ninety minutes, as they planed steadily down from high altitude, Fred Noonan and Amelia Earhart scanned the horizon for a sight of land, calling the *Itasca* at regular intervals. They made no further contact in that time. Then, at 7.42, the operators on the cutter shouted with relief at the sound of Amelia Earhart's voice.

'We must be right on top of you, but we can't see you. Our gas is running low. Have been unable to reach you by radio. We are flying at an altitude of a thousand feet. Please take a bearing.'

Once again the operators swivelled their direction-finding gear, hurriedly now, knowing that an accurate bearing or course to steer was urgently required. But it was not easy to get a sharp minimum on voice transmissions even under good conditions. The heavy static interference that still crashed into their ears made it impossible.

Just before eight o'clock, their estimated time of arrival, Amelia Earhart began a wide circuit, searching for the speck of land that must surely be near. For 22,000 miles their flying and navigation had been accurate. Why should anything have gone wrong on this critical leg, when they had both taken particular care? She decided to try to use her radio compass on transmissions from the *Itasca*; morse signals would not be so badly affected by the static. She called them again.

'We are circling but cannot hear you. Go ahead on 7,500 kilocycles, either now or on the scheduled half-hourly time.'

She tuned her receiver to the new frequency while the operators on the cutter began transmitting their call-sign in morse code, interspersed with long dashes which were meant to give a firm reading on the compass needles. Still it was impossible to get a clearly defined bearing.

'We are receiving your signals,' said Amelia Earhart, 'but are unable to get a minimum. Please take a bearing on us and answer with voice on 3,105.'

But the brief interlude of partially improved reception had ended. Further transmissions were suffocated by static. For the next 45 minutes the crew of the Elektra, imprisoned in their tiny cockpit, their tenuous contact with the outside world broken and their fuel running low, continued their desperate search for Howland Island, for land, for any land at all. Then the static suddenly parted like a curtain and Amelia Earhart's voice broke through.

'We are in a line of position 157–337. Will repeat this message on 6,210 kilocycles. We are running north and south. We have only half an hour's fuel left and we cannot see land.'

The operators on the cutter switched their receivers at once to the new waveband, but the ether was silent and they heard no more. Amelia Earhart's voice had been calm and her message had been phrased undramatically, but there was no mistaking the desperate plight she was in. The operators called her again and again; the men on deck scanned the sky. Still nothing was heard or seen, and the minutes ticked by. After an hour had passed, then two hours, and then three, the moment came when even the most dedicated radio man realized that there could be no reply to his calls. The Elektra could be airborne no longer; somewhere in the vast Pacific Ocean, Amelia Earhart and her companion must have come down. It was possible that they were hundreds of miles away and had found an inhabited island. They might be clinging to a coral reef. But it was far more likely that they had come down in the sea.

Although electric storms were building up in the area, as foreshadowed by the heavy static, the waters around Howland Island were smooth and the sky was clear. The plane carried a rubber lifeboat, lifebelts, flares, a signalling pistol, emergency rations and water. The Lockheed Company believed that with the petrol tanks empty and the cocks closed the plane might float almost indefinitely on a calm sea. There was every reason to believe that the two fliers were alive. But the difficulty was to get help to them quickly. They were in an unknown position in a remote area, and it would be days before a thorough search could be made.

A British cargo steamer, the *Moorby*, bound for Sydney, was diverted to the area to search for wreckage. The battleship *Colorado*, carrying three catapult planes, left Pearl Harbour; but that was 2,000 miles away. The aircraft carrier *Lexington*, with a complement of 62 planes and accompanied by six destroyers, put to sea from San Diego, California; but that was more than double the distance. Only the *Itasca* was on the spot.

It sailed at once on a systematic search south of Howland Island.

Nothing was found. There was not even any very precise indication where to look. Then, with the naval search units still hundreds of miles from the scene, came the first post-flight message from the missing plane.

'Can't last much longer. Plane is sinking.'

Three naval operators at Honolulu heard the message, in morse code, garbled and inexpertly sent, about 24 hours after the plane was assumed to have come down. All along the Pacific coast a chain of radio amateurs, headphones clamped to their ears, listened for further transmissions. Hour by hour, ragged, erratic, indecipherable signals kept hundreds of listeners concentrating on their receivers, until at last came a further coherent message, garbled enough but making some sense.

'281 north Howland ... call KHAQQ ... don't hold with us much longer ... above water ... shut off.'

KHAQQ was the Lockheed's call-sign. It looked as though the two stranded fliers were trying to indicate their position somewhere north of Howland Island; the figures might refer to mileage or direction. As a clue it was flimsy enough, but it did give the ships and planes converging on Howland Island something to go on. Everyone had assumed that the plane must have come down short of its destination; now it was realized that it was much more likely that it had overshot.

There was one puzzling feature. How could wireless messages be transmitted from the plane if it was on the water? Technicians said it was impossible. But a fragment of a further message seemed to resolve this doubt. 'We are on a coral reef just below the Equator ... don't know how long we can last ... we are OK but a little weak.' Radio men agreed that Noonan might somehow have rigged up the wireless gear if the Lockheed was marooned on a coral atoll.

Back in California, G. P. Putnam and Mary Noonan, Fred Noonan's wife, watched the ticker tapes endlessly in a San Francisco newspaper office, across the bay from Oakland airport, where they had hoped to greet the fliers at the end of their flight. Putnam caused messages to be sent every fifteen minutes by a powerful radio transmitter. 'Attention Amelia Earhart. Your signals have been heard. Help is on the way.'

Help was indeed on the way. The approaches to the area were swarming with craft. But squalls and the threat of equatorial rains promised to make searching difficult. Rescue planes from Honolulu were driven back by storms. And across Howland Island swept violent winds which brought lightning, snow and sleet. The carrier planes were forced by severe icing and sluggish controls to abandon their searches and land back on the carrier decks. Worst of all, it was clear that no floating plane could live in such weather; even on a coral atoll it seemed doubtful if anyone could survive for long.

Yet, when the storm was over, the messages continued. Many people had doubted their authenticity from the start. Now, as they became more and more garbled and contradictory, the truth became starkly apparent. All the post-flight messages had almost certainly been the work of the warped and inhuman minds of hoaxers. Fighting their disgust and anger, the searchers went on. A thorough enquiry was begun on Canton Island, 400 miles to the south-east, and on all the islands of the Phoenix group, but no one had seen or heard anything of the missing plane and nothing was found. At last, on 18th July, the search was called off. It had occupied 4,000 men for over a fortnight, cost an estimated million dollars, and covered an area of a quarter of a million square miles, all to no purpose.

The frank, artless personality of Amelia Earhart had enabled her to escape controversy in her lifetime. But now, with relentless energy, it pursued her memory. Within a few months a startling rumour swept America that the two fliers had been on a secret mission for the United States Government. They had deliberately flown off their course, went the story, to spy on Japanese war preparations in the Pacific. Probably the Japs had captured them and were holding them prisoner. The Japs denied it. After a year or so the rumours became submerged in the more tangible tragedies of war.

When hostilities were over, many people felt it was time that the rumour was thoroughly probed. All former Jap-held territory in the Pacific had been captured. Surely, if the evidence was there, a little digging would uncover it. Suspicions had been stimulated by a photograph found on a dead Japanese soldier at Okinawa. It was clearly a picture of Amelia Earhart and her plane. Then, on the island of Saipan, in the Marianas, a whole album of similar photographs was discovered.

It was known that in 1937 the Japs had started building a small airstrip on Saipan, in violation of existing treaties. Had Amelia Earhart and Fred Noonan stumbled on this secret and paid for it with their lives? Or could they, perhaps, still be alive, prisoners on some uncharted atoll? Efforts to solve the mystery of their disappearance were redoubled, yet in spite of the most comprehensive enquiries the United States Intelligence staffs were unable to find any evidence that Amelia Earhart had been seen or captured by the Japs. No record of any such incident was found in any captured Japanese document. A former Japanese naval captain stationed at Saipan at the time was thoroughly interrogated but denied all knowledge of Japanese implication. Many records, though, had been destroyed, and many of the people who might have been expected to know the truth were dead. Eventually the Americans had to admit defeat, and enquiries were allowed to lapse. It was not until 1960 that they were taken up again.

In that year a team of reporters from the Columbia Broadcasting System arrived in Saipan to gather material for a radio programme about Amelia Earhart. The islanders could give only a vague testimony, but several claimed to have seen a white woman and a man arrested after their plane had come down in the bay. Some said they thought the prisoners had been executed. Then, in remarkable corroboration, a Japanese-born woman working in California came forward with a more detailed story. As a young girl of eleven, she had been riding her bicycle towards the harbour at Saipan, taking a packed lunch to her brother-in-law, who worked in the docks. 'In the distance, on the beach,' she said, 'was an aeroplane. About a hundred feet away were this man and woman. The people kept saying "American lady pilot".

'They stood there talking to some soldiers, and then they were led away into a car. I never saw them again.'

Who else could this have been but Fred Noonan and Amelia Earhart? All that remained to clinch matters was to find some trace of the plane. Even after 23 years there might still be a chance of finding a few identifiable pieces of wreckage in the harbour. Dragging operations were begun, and sure enough scraps of an old plane were found, including an electric generator, its serial number still preserved beneath the slime and rust. Details were cabled to the Bendix company, who had equipped the plane. Their reply was frustrating: this was not the generator originally built into Amelia Earhart's Lockheed. But the suggestion was put forward that, at some stage of the 22,000 miles of safely completed flight, the generator might have broken down and been replaced. This was feasible, and enquiries went on.

Eighteen months later, in November 1961, the last piece of the jig-saw puzzle seemed to fall into place. The remains of two unidentified bodies were found in a shallow, unmarked grave in Saipan. The skeletons were flown to California, where Dr Theodore McKown, an anthropologist at the state university, made a thorough analysis. He was not long in pronouncing that the bones and teeth were of oriental origin. They could not possibly be the remains of the missing fliers.

With this last shattering failure, enquiries ended. And working back through the material clues, one by one they can be shown to be dubious or false. The skeletons didn't fit. The generator didn't fit. The photograph found on the dead Japanese soldier proved on examination to have been taken before the plane left Lae. Like the album, it had been captured when the Japs took New Guinea. It was odd, too, that witnesses should come forward in 1960 to describe and corroborate the alleged landing off Saipan when none had apparently done so during the exhaustive enquiries fifteen years earlier, when memories must have been fresher.

In an attempt to explain how the Lockheed got so far off course as to reach Saipan, it has been hinted that Amelia Earhart and Fred Noonan, hurrying to reach America in time for the 4th July celebrations, took off physically tired and with unserviceable chronometers from Lae. The suggestion, which is not borne out by the known facts, is out of tune with the care and prudence in which the rest of the flight was undertaken and does the fliers a gross injustice. To assess the likelihood of the plane's straying thousands of miles off its course, as it would have to do to reach Saipan (which is 1,800 miles due north of Lae), or any other Japanese-occupied island, one needs to know more about that shadowy figure Fred Noonan. Fortyish, married only a month before the flight, he had been navigating ships and planes for nearly twenty years. A master mariner, torpedoed three times with the British Navy in the First World War, he had transferred to Pan American Airways in the twenties and was a doyen among Pacific fliers, one of the finest aerial navigators in the world. That is why Amelia Earhart chose him to accompany her.

Both Noonan and Amelia Earhart knew that the flight from Lae to Howland Island was the toughest part of the whole trip. It was not a journey they were taking any chances on. The notion that Noonan could have been thousands of miles out in his calculations when the plane was due to arrive at Howland can surely be discounted. It is not likely that he was more than a hundred miles adrift. But this was more than enough to miss Howland Island.

It is easy to imagine the two fliers in those last hours, while their fuel lasted, and while the men of the *Itasca* listened and waited and called, searching for their haven in vain, often coming close to it but never seeing anything more tangible than the shadow of a cloud on the water. It is even possible that before their fuel gave out they made a safe ditching: they may have floated and kept alive for a day or so. Then came the storm. By the time the searching planes from the *Colorado* and the *Lexington* reached the area, all trace of the plane and its occupants would have been swept away.

It is, perhaps, even more likely that the two fliers would

have gone on searching until the last possible moment – until the engines of their Lockheed coughed and finally cut. If that is so, it is very probable that the end came then.

'When I go out,' Amelia Earhart had once said, 'I should like to go quickly, in my own plane.'

The chances are that she did.

2

THE *LADY BE GOOD*

THE men of the oil-prospecting team, staring out of the cabin of the chartered Dakota at the lifeless wastes of the Sahara, had not seen a trace of human habitation or movement for many hours. Leaving behind them the Libyan coastal strip, with its crowded Arab towns and its litter of abandoned trucks and planes left over from the war, they had aimed for the heart of the Libyan Desert, where the area of the proposed concession lay. At length, after overflying the treacherous dunes of the Libyan Sand Sea, they emerged into the core of the concession, the unknown, infinite, inaccessible void of the *serir* of Calanshio, a flat, gravel plain hard as granite whose northernmost point was some 350 miles south of Benghazi. Now, as their practised eyes evaluated the soil and rock stratum, they savoured the thrill of exploration. It seemed unlikely that Western man had ever before penetrated into this trackless plain.

Here, in a vast furnace shaped like an inverted U, never before penetrated by geologists, unexplored and unmapped, rarely traversed even by the Sahara's own nomadic tribes, they entered a blind spot of God's creation, a divine aberration, thousands of square miles of rock-strewn desert constituting the most arid spot on earth. It was here, right in the middle of the sterile plateau, that they saw the wrecked plane.

For a moment, in that shimmering heat, it seemed to move forward, as though in the act of take-off. Its thick, bulbous nose was pointing towards them. Then they noticed the downward slant of the port wing, and the absence of the starboard outer engine; the plane had evidently been damanged in a belly-landing. The other three engines were intact, their propellers bent but erect, like antlers. Flying low over the wreck, the men in the Dakota cabin saw that the front part of the fuselage was pressed hard against the ground, in the attitude of an animal that had broken its back. They saw, too, the marks made by the plane in its landing, showing how the plane had struck the ground in a shallow dive and slewed round drunkenly, finishing up in the opposite direction to its line of flight.

The pilot of the Dakota circled the scene for some minutes, hypnotized by the strange anachronism below him. He believed he was looking at the wreck of a B24 Liberator bomber – but he could not imagine how it had got where it was. There was no sign of the crew. When he landed back in the Libyan coastal strip at the end of his survey flight – it was November 1958 – the pilot reported his sighting to the United States Air Force at their base at Wheelus Field, near Tripoli. But it stimulated little interest.

'We've lost no ships down there,' said the Americans. 'And as for a B24, they've been obsolete in this theatre for fifteen years. Even during the war they never went far inland. Sorry, Bud. It can't be one of ours.'

The mystery of the stranded Liberator – if it was a mystery – thus remained unsolved. The only action taken was that the pilot marked the site of the wreck on the maps that would be used by the motorized oil-prospecting teams when they made their painful way across that oven-hot surface. It was another year before the teams set out. Then, after forcing their way through the undulating dunes of the Calanshio Sand Sea, they emerged at last on to the barren, sun-drenched plain.

Hour by hour the stifling stillness of an area that could not succour any living thing was broken only by the whine of six-cylinder engines and the crunch of heavy-duty tyres. There were three vehicles in the convoy, and like hermit crabs they crawled purposefully forward, their occupants pausing at intervals to dig with mechanical claws into the parched, rubble-strewn soil. For these men the excitement of treading where it seemed that no man had stepped before was tempered by a sense of trespass. Steadily they drove southward towards the one landmark they could expect to find on that vast featureless plain – the wreck of the Liberator, marked on their maps.

It was just after midday when they saw it. Soon they drew level, and as the vehicles halted men jumped down and walked round the wreck, warily at first, as though uneasy at what they might find. The silver panels of the fuselage and main-plane, polished and rustless, glinted in the sun. There was no erosion, and very little dust. The wreck must be comparatively recent. They knew the Americans had disclaimed any knowledge of it. Perhaps Liberators had been used by some emergent African nation. But then they saw the American white star on a blue background clearly etched on the fuselage; it was an American plane after all. Other identifying marks were the number '64' painted on the nose – and the plane had also been given a name. The survey team recognized it as the title of a Gershwin song. The plane had been christened the *Lady Be Good.*

Stepping forward, the survey men peered inside the fuselage. The heat inside was breathtaking. There were no bodies, no tattered uniforms, no evidence that anyone had been hurt in the crash. Everything seemed ready for flight. Flying-jackets hung from protruding stanchions of metal, urns spouted water when taps were turned, there was a flask of drinkable coffee. Wherever the crew had gone, they had evidently left in a hurry.

The members of the survey team pushed their way forward into the front crew-compartment. Here they entered an eerie hothouse, manned only by ghosts. The cockpit was undamaged and the instruments stared back blankly, revealing nothing. Somebody spotted the navigator's log. And this document put the date of the plane's arrival in this desolate place beyond all doubt. It told of a bombing mission to Naples harbour; the date was 4th April 1943. The Liberator *Lady Be Good* had lain in crippled agony on this inaccessible plateau for more than 16 years. What had happened to her crew?

With the information provided by the survey men on their return to base, the first part of this absorbing flashback was not difficult to reconstruct. The plane was now positively identified as having belonged to the 376th Bomber Group of the US Air Force, which in April 1943 had been based at Soluch, 34 miles south of Benghazi on the Libyan coastal strip. The pilot for this raid on Naples harbour had been First Lieutenant William J. Hatton, of New York.

Hatton and his crew had been Yankees to a man. Their home towns were stretched across the northern states from Massachusetts through New York, Pennsylvania, Ohio and Michigan to Illinois. Co-pilot was the fatalistic Robert Toner, who did not believe that in war man's thoughts should dwell on the future. Navigator Hays, shortest man in the crew, with a receding hairline, was the comedian. 'Rip' Ripslinger, the engineer, and Guy Shelley, gunner, were the giants. The other crew-members were John Woravka, bombardier, Vernon Moore, radio operator, and Robert LaMotte and Samuel Adams, gunners. Between them they represented a sizeable slice of Yankee civilization.

Slowly, from US Air Force records, the story of their last flight began to take shape. After flying across to North Africa in March 1943, they had been allowed a few days to acclimatize before being sent on their first mission. This was on Sunday, 4th April, against Naples harbour. The force assigned to the task consisted of 25 unescorted B24 bombers from Soluch, split into two sections. The plan was to strike at the target from high altitude at last light, then to break up in the gathering darkness and fly back singly at reduced altitude to North Africa.

A blinding sandstorm swept across the landing-ground just before 1.30 that afternoon as the first section of 12 planes prepared to take off. One plane soon returned with engine trouble aggravated by the sandstorm, but the others went on, bombed the target, and returned safely to base. The second section, taking off a few minutes later, were caught in the worst of the sandstorm, and they did not fare so well. Seven of the 13 planes in this formation turned back with engine and other troubles. In two of the remaining six, the oxygen masks of the waist gunners froze at high altitude and the gunners blacked out. In diving to revive the gunners, the pilots lost formation and turned for home. That left only four out of the original 13. Both the leader and the deputy leader of the section had fallen out. It was a brand-new B24 which now assumed the lead, with a brand-new crew. On her bulbous nose was painted the title of a Gershwin song.

The four surviving planes reached Sorrento, 30 miles south of Naples, at 7.50 that evening. Even at their 25,000 feet altitude the sun had disappeared and it was almost dark. Liberator crews were accustomed to bombing in daylight; it was too dark, in the opinion of the four pilots, to risk bombing this particular target. Italy was on the brink of switching sides, and the crews had been give strict orders not to risk the needless slaughter of civilians. The harbour of Naples was too near the crowded city for safety, and led by Hatton the four crews broke formation, turned south and headed back for Soluch. Two of them bombed Catania airfield, in Sicily, on their way home. The other two jettisoned their bombs in the Mediterranean.

Of these four planes, one ran out of fuel and landed at Malta, two others landed safely at Soluch, one at 10.45 and the other at 11.10. That left only one plane – Hatton's – unaccounted for.

The formation had been under orders to maintain absolute radio silence except in emergency. But at twelve minutes past midnight, with his fuel seriously low and the North African coast not yet in sight, Hatton called Benghazi.

'Benina Tower from Six Four. Do you read? Over.'

'Five-by-five. Over.'

'Request an inbound bearing. Over.'

Hatton continued to press his transmitter button while the ground operator swung his direction finder.

'Six Four from Benina Tower. Your bearing from us is three three zero.'

'Roger three three zero. Out.'

This put the *Lady Be Good* on a direct line between Naples and Benghazi. It seemed that they were on course after all. But one fatal detail had been overlooked. A single American radion direction finding station of April 1943 was unable to distinguish between a true bearing and a reciprocal. Thus it had no check on whether an aircraft was inbound or outbound. Hatton anyway had asked for an inbound bearing, and the ground operators naturally assumed that he was approaching Benghazi from the sea. But for all they really knew – had they thought about it – Hatton could have been approaching on the line of direction exactly opposite to that indicated by the bearing. He could have been approaching from inland. More significantly still, he could have overflown Benghazi already and be actually *heading* inland.

The sandstorms of the day had left a residue of haze which partially obscured the coastline, spilling several miles out to sea. Hatton and his crew had crossed the coast without realizing it. The deception had been aided by the treacherous *ghibli,* the wind which had followed them on their way to Naples and then turned cunningly with the formation and followed them back. Thus they had reached the North African coast before they had alerted themselves to a close downward scrutiny. And they had missed the collapse of the radio compass needles when overhead Benghazi.

Because of the order for radio silence, Hatton did not call Benghazi again for some time. When he did he got no reply. The Liberator had passed out of range. Each minute, as they sat there patiently, their four engines took them further away. Looking down for some point of recognition, some sign of a coastline, a glow of light or the lace fringe of the sea, they saw only the undulating dunes of the Libyan Sand Sea. These dunes exactly imitated the gentle swell of the calm Mediterranean. Still looking for the coastline, they flew on.

On the ground at Soluch the overflying plane had been heard. But there were many planes in the area, and many airfields. There was nothing to connect this plane with the 376th Bomber Group. The only clue, a flimsy, circumstantial one, lay undetected in the de-briefing files at Soluch. A pilot in Hatton's formation, describing the approach to the target had reported that he 'thought the leader was Number 64'. If that were so – if Hatton had been out in front at that stage – why should he be so late back? The timing of his request for a bearing – more than an hour after the planes which had been formating on him had landed – could just possibly have led someone to the correct conclusion. That it did not do so is not in the least surprising and reflects no discredit on anyone.

Shortly after half-past one they passed clear of the sand dunes and began to overfly a flat plateau 500 feet high, a hundred miles in length and a hundred miles wide, 300 miles south of Benghazi. This was the barren *serir* of Calanshio. At first, perhaps, it suggested to them nothing more than a smooth stretch of sea, possibly in the protected Gulf of Sirte. They must surely make a landfall soon. But somewhere over this featureless plain it may be that Hatton and his crew recognized, at first with relief and then with dismay, that they were flying over Africa.

There came the time, as they gazed downwards and circled, perhaps, in a despairing effort to orientate themselves, when the engines coughed and spluttered for lack of fuel. One by one Hatton closed the throttles until, flying on a single engine and losing height, he ordered the crew to jump.

Had they believed they were still over the sea, it seems likely that Hatton would have attempted to ditch. To bale out into the sea, even into the Mediterranean, meant almost certain death from exposure within a few hours. They had no idea where they were and for a long time their signals had gone unanswered. A ditching meant all the advantages of crew dinghies, an emergency 'distress' radio, drinking water, rations and survival equipment. So it seems almost certain that they now realized they were over land.

'That's about the end of the gas, fellers. Bale out. Bale out.'

Had he but known it, the ground beneath him might have been specially constructed as one vast airfield, admirably suited to a wheels-down landing. They could have had all the advantages of a ditching, plus the protection offered by the plane. But now Hatton trimmed the Liberator to hold its gentle glide before leaving his seat to follow the rest of the crew. As they pirouetted to earth they saw the plane gliding steadily on, losing height more steeply as it disappeared to the south.

At this point, with an unpleasant jolt, the first flashback

ended. This much reconstruction had been possible from a study of US Air Force records and a close scrutiny of the plane. It was also established that the crew had been posted missing after the Naples raid, the assumption being that they had come down in the sea. A routine search along the coast and out to sea had proved fruitless, and within twelve months the deaths of the crew had been presumed. What had actually happened to them had always been a mystery, and even now, although their plane was found and identified, their fate was unknown.

Weeks of conscientious searching by ground and air parties, including helicopters, failed to solve the riddle, although from marker triangles found on the *serir* it was clear that the crew had tried to make their way north. Then, 25 miles short of the Sand Sea, the tracks abruptly ended. After four months the search was abandoned. It seemed that the desert was to hold its secret. But in February 1960 came another discovery by oil prospectors which provided a second long and gripping flashback. This discovery, which included a diary, traced the story from the moment of the parachute jump.

By operating hand-held signal flares and firing their revolvers, eight of Hatton's crew had managed to join up at dawn. The ninth man, John Woravka, the bomb-aimer, was missing. The others searched and called to him for two hours in vain. Perhaps something had gone wrong with his parachute. Or perhaps he had landed further north – he had been one of the first to jump.

Because of the controlled manner in which the B24 had flown on, these men escaped the traditional dilemma of forced-landings in the desert – whether to stay with the plane or to start walking. They had no idea how far south the plane might be. Presumably when it crashed it must have disintegrated. Nothing in it would be of any use to them. They had no alternative but to start walking northwards for the coast. They did not know that the B24 had effected a very respectable robot belly-landing twenty-five miles to the south and that all the equipment in it was perfectly preserved. Neither did they

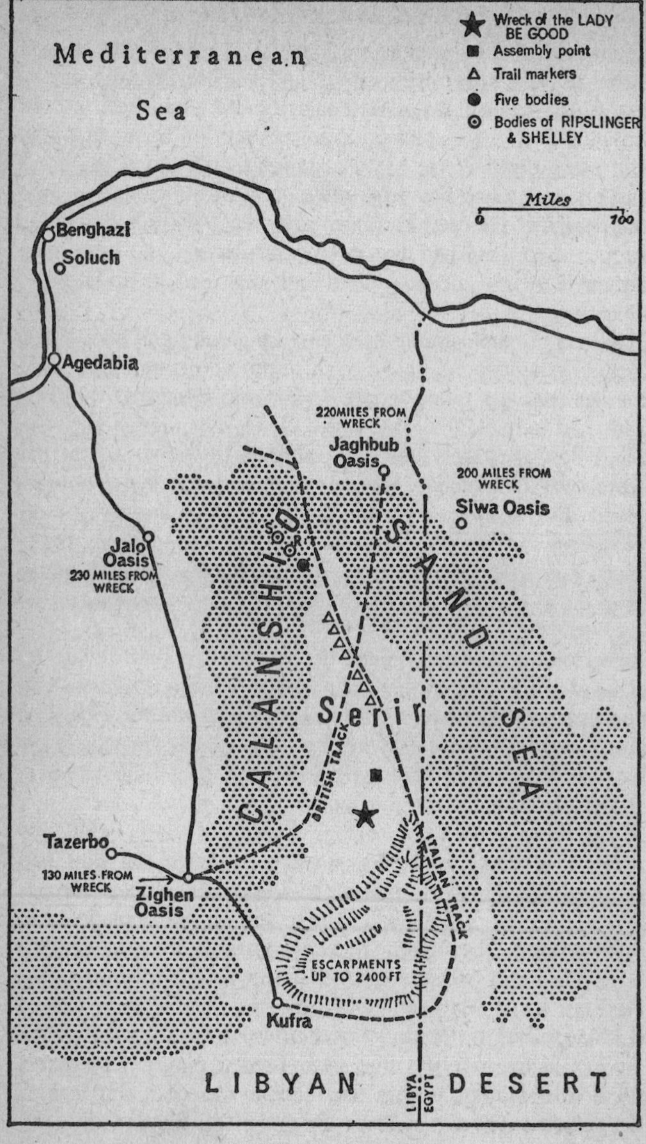

2. The Serir of Calanshio

know that they were more than 400 miles south of Benghazi and 300 miles from the nearest point on the coastline, right in the middle of the inaccessible *serir*.

Ninety miles ahead of them, and 50 miles to either side, lay the impenetrable dunes of the Calanshio Sand Sea. Fifty miles behind them stretched a mountain range of rocky outcrop that was equally impassable. They were incarcerated in a vast natural prison which, but for some external miracle, must become for them nothing but a crematorium and mausoleum. Even possession of the equipment and rations preserved in the plane could only prolong what promised to be an excruciating agony. Mercifully, they were unaware of their situation.

They imagined that they might be 50 miles inland, perhaps even a hundred. Early that morning, Monday, 5th April, they started to walk northwards towards the coast, spreading out in the hope of finding Woravka. With them they carried a single flask of water, a few rations, and the canopies of their parachutes, which they severed from the rigging lines. These canopies would serve as protection against the cold nights, as markers, and to catch water if it rained. They had no means of knowing that there was no dew in this area and that not a drop of rain had fallen for many years.

They decided to limit themselves to one mouthful of water per day, measured out in the metal cap of the flask. It held little more than a teaspoonful. A second flask had been carried by Woravka, and his absence meant that their meagre water supply had been halved at one go.

The sun was an unrelenting heater poised immediately above their heads. Yet the north-westerly breeze that had blown them back across the Mediterranean was just enough at ground level to cool their faces. They made good progress at first across the plateau. Most of the time they scanned the sky for a glimpse of searching planes. It occurred to them that the most likely thing to be seen from the air was the wreck of the B24, wherever it was; searchers would try to pick up the trail from there. They stopped at regular intervals to gather some of the larger stones that littered the plain, each time laying out a ground marker in the form of a large triangle, its apex pointing north-west. They tore strips from a parachute and weighted them down with the stones, to make the markers stand out from the air.

The plain that seemed so trackless had in fact been traversed at least twice before – by military vehicles. An Italian track came up from the Kufra Oasis to the south through a mountain pass before heading north-west across the plateau, in the direction of Benghazi. A British track ploughed through the Sand Sea from the El Zighen Oasis, 120 miles to the south-west, then headed obliquely across the plain for the Jaghbub Oasis, 200 miles to the north-east. The tracks crossed about halfway up the plateau. By great good fortune the crew had parachuted down within a few miles of the Italian track, in the southern half of the plateau. As they spread out to look for Woravka it was inevitable that they should find it. Having found it, it was equally inevitable that they should follow it. It led exactly in the direction they wanted to go.

Consulting their escape maps, they found that there were in fact three desert tracks leading coastwards, one of which clearly headed for Benghazi. This must be the one. It did not occur to them that they were so far inland that their maps, which covered only the coastal fringe up to about 200 miles inland, were useless to them.

When night fell they covered themselves with their parachutes and tried to sleep. But the temperatures knew no moderation. The liquefying heat of the day changed abruptly to a refrigerating cold, with a temperature only a few degrees above freezing. They huddled together but could not sleep. Eventually they abandoned the attempt, packed their parachutes again and started to walk, resting occasionally but not trying to sleep. Next morning, as they plodded on steadily, their track crossed a second, much wider track which came up behind them from the south-west. This was the British track, heading north-east for the oases of Jaghbub and Siwa. The two tracks lay before them like open scissors, stretching away into the unknown distance. The choice of which one to follow might hold the difference between survival and death. Yet there was a natural, corporate reluctance to change the course they had set. As a compromise, two men, Hays and Adams, volunteered to follow the new track for a while to see where it led. But soon, discouraged at finding nothing and afraid of losing touch with the main party, they traversed across the intervening ground and back to their comrades on the original track. That at least pointed in the direction of Soluch and Benghazi.

The north-westerly breeze that had mitigated the heat of the previous day had gone, leaving them exposed to a fiery orb that seemed to descend upon them out of the heavens and stare them in the face. Temperatures which rose to more than 130 degrees in the shade were meaningless in an open-air hypocaust where shade was unknown. By noon the devouring flame of the sun had forced them to stop. Even the air itself seemed to burn, with all the ferocity of a highly inflammable fuel.

Most of the men were without sunglasses and they were forced to screw their eyes up or shield them with an arm or a hand. Their eyeballs were stricken with the unremitting dazzle of the sun and seared by its heat. But not even the flesh and bone of an arm could wholly protect them from that blazing light. And when they turned their heads to avoid it, the reflected light boiled up at them out of the plain. Even the stones were the cinders of an enduring fire.

During that interminable afternoon the anguished depths of human suffering were revealed to all of them. There was nothing left for them to fear. Together, no doubt largely through their own shortcomings and inexperience, they had fallen into this boundless incinerator. Together they took refuge in prayer.

There was still no sign of a searching plane, but they continued to hope. Help must surely be coming. Pray God it come soon. Meanwhile they did their best to help themselves. When the sun weakened they laid out another marker triangle before resuming their north-westerly trek. Walking for fifteen minutes, resting for five, they kept going throughout the night. Next morning, Wednesday, 7th April, they struggled to keep the same routine. But now they could not only feel their own individual weakness; they could see it in each other.

They began to realize that they couldn't get much farther. One or two more afternoons in this cauldron of heat would finish them.

They still rationed themselves to a cap-full of water a day. Half the water had gone. They still laid out ground markers at regular intervals along the track. They prayed continually. That afternoon they again suffered the tortures of hell.

When dusk came they tried to rest and regain their strength, but their thirst was too insistent and their bones too sore for sleep. They began walking. With regular intervals for rest, they still covered the ground well. Ahead of them next morning they sighted with triumph the dunes of the Calanshio Sand Sea. In an area where, without adequate water, the limit of distance that a human being might cover was thought to be 20 to 30 miles, they had walked for about a hundred miles diagonally across the hostile plain.

They had no idea what lay behind those dunes. Were they a sign that they were nearing the coast? At first they must have thought so. But soon the difficulty of making progress through this soft, shifting sand deepened their misery. There was no sign now of any motor track; it had been obliterated by those shifting sands. Except where they could follow the line of the dunes, they sank knee-deep into the treacherous sand. Even the camels of the nomads could not safely cross these dunes. Somewhere in this wilderness were the bleached skeletons of several that had attempted it.

During the day the breeze returned, but it brought with it a cloudy suspension of blown sand. The terrible pain in their eyes was exacerbated almost beyond endurance by the fine grains of sand that seeped under their eyelids. LaMotte was now almost completely blinded, and he tramped forward unseeingly. The extinguishing of that searing light brought no relief to his pain, but neither did it quench his spirit. The weakness of all of them was now pitiful. Adams and Moore were dropping behind, their strength almost gone. Yet still these eight men kept together as a party, making slow and uneven progress to the north-west. Everything was subordinated to this laboured progress. They had no spare strength now to plant markets for the air search that they still hoped and prayed for together.

Next day, Friday, 9th April, only three men had the strength to go on. They were Guy Shelley, 'Rip' Ripslinger, the biggest men in the crew, and Vernon Moore, who on the previous day had been one of the weakest. The others lay exhausted in the sand, seeking nothing more than the final release of death. It was agreed, in an eight-man conference conducted in little more than a whisper, that Shelley, Ripslinger and Moore should go on if they could to look for help. The larger party would retain the flask. In any case it was almost empty.

At dawn next morning the five men who still lay collapsed in the sand watched the remaining three set off into a cool north wind. Once again the day was a hideous torture from which the five exhausted men prayed for deliverance, either by rescue or death. At night their emaciated bodies froze in the bitter cold. They had only one parachute left between them. The others had been used to make the markers, or had been dropped in the sand by men too weary to carry them further.

Next morning the five men prayed again in unison, and watched the sky endlessly, wondering how the other three were faring, whether they had managed to get through the dunes, what they had found beyond them if they had. That morning they saw the first living things they had seen for a week – a pair of migrating birds in transit; it brought them tantalizingly close to the idea of a continuing existence.

They were so weak now that they could hardly struggle to their feet, and when they tried to walk they staggered and fell. Every bone in their bodies ached with its own separate intensity. In the merciless heat of the afternoon they prayed again for death. In the evening they drank practically the last of the water.

When they looked at each other next morning they were shocked to find that already they were little more than skeletons. Yet they still felt that if they could reach water they would be able to go on. There was barely enough left in the bottom of the flask to moisten their tongues.

The fatalistic Robert Toner still made daily entries in his diary. But the time inevitably came when the muscles of his hand and wrist would no longer answer his brain and will. His last stoical words, on Sunday the 11th and finally on Monday the 12th, eight days after the drop, were of their continuing hopes of rescue. And of the cold nights. Fleshless as they now were, it was not the heat of the day that conquered them in the end, but the long nights of exposure to the intense cold.

The five bodies were eventually found by British oil prospectors on an upward slope of the sand dunes, as though they had fallen in a final struggle to see over the top of the next hill. At once the Americans began a new search for the rest of the crew.

It seemed that final elucidation of the 17-year-old mystery was near. How far had the advance party got? Had they kept together? Could they, perhaps, have reached some half-forgotten desert track? Could any of them possibly be alive, captured, perhaps, by nomad tribes? But once again the most intensive and highly organized search failed to find any trace of Shelley, Ripslinger and Moore. Or of Woravka. And again, months after this second search was abandoned, it was an oil-prospecting team who came upon two of the bodies of the advance party.

It was plain that as one man had dropped the others had gone on, until the last hope for all of them lay with one man. Then he, too, finally collapsed into the sand. Almost certainly this was Guy Shelley. Ripslinger had somehow managed to penetrate 20 miles into the dunes. Shelley was found seven miles further on. Every one of those last seven miles must have been an epic of endurance. When he finally collapsed he had covered about a hundred and forty miles, virtually without water; but he was still 85 miles from the Jalo Oasis. Like the others, both Shelley and Ripslinger were found on upward slopes. Moore's body was never found. He had been the weakest of the three, and was probably the first to fall out.

The terrible irony of that tragic march across the furnace of Calanshio was now distressingly clear. All eight men, and most of all Shelley and Ripslinger, had achieved a feat of tenacity and endurance which might, if their escape maps had covered the area into which they fell, have saved all their lives. For had they set out from their assembly point in a south-westerly direction, towards the oasis of El Zighen, the distance they would have had to cover would have been very little more than that achieved by Shelley. And more significant still, they would have passed their plane on the way, finding the precious urns which 17 years later still held uncontaminated, drinkable water. They would surely have reached El Zighen.

Only one secret remained, and even that was later resolved by the finding of the body of the ninth crew-member, parachute unopened and embedded in the sand, where he fell that night from the plane. This was the body of John Woravka, the man to whom God was merciful when he jumped from the *Lady Be Good*.

3

THE ATLANTIC PRINCESS

THE dinner party at the Savoy had been a glittering affair. Cappone, the Italian restaurant manager, had fussed over the diners in his aloof, elegant manner. Latry, the chef, had surpassed himself. In the subdued light shed by the chandeliers they had danced the Charleston, the black bottom, the turkey-trot and the fox-trot, to the music of Carroll Gibbons and the Orpheans. They had chuckled at the thick-lipped, burnt-cork drolleries of the Two Black Crows. The year was 1927.

Seated at the table were two of the most experienced fliers of the day. First was the tall, sunburnt, military-moustached Irishman Colonel Freddie Minchin, better known as Dan Minchin, a leading bomber pilot of the First World War, with the DSO, MC and Bar, now pioneering the long-distance routes for Imperial Airways. As a young man he had been dubbed 'the longest, blackest and slackest thing in Kildare', and he was still capable of letting his hair down at a party; but he was always meticulous when on parade. Over the years he had built up a solid reputation for single-mindedness and a quiet, sometimes melancholy charm.

The second man, Captain Leslie Hamilton, shorter than Minchin, was dark-haired, pale-faced and extremely good-looking, with the reputation of being irresistible to women. Like Minchin he was married, but his wife was in New York and there was talk of an estrangement. Hamilton had caught the public imagination as the 'aerial gipsy' because of his wandering life as a private commercial flier, in the course of which he frequented most of the Continental glamour-spots. He, too, had been a distinguished wartime and post-war flier in the RAF.

But the central figure at the dinner party was a slim, graceful woman of indeterminate age, her face unlined and her hair unflecked with grey, who talked with easy assurance, even authority, and who was herself a flier of long experience. She was the hostess, and the dinner party was her idea. She was the Princess Ludwig Loewenstein-Wertheim, sister of the Earl of Wexborough and better known in aviation circles as Lady Anne Savile. Her husband, a German prince, had been killed in the Philippines in 1899, two years after she married him, fighting for the Spaniards against the Americans, and after the First World War she had regained her British nationality. Under her expertly-applied make-up her face retained much of its youthful beauty. Her small features and smooth skin had not coarsened at all. She looked, perhaps, 45. In fact, she was 60.

She had invited these two well-known fliers to the Savoy to put a proposition to them. She was prepared, she told them, to finance a project which she hoped would capture the imagination of the world, bring fame to those who accomplished it, and prestige to their nation. What she had in mind, she said, was an Atlantic flight from east to west. The Atlantic had never been flown in that direction. All successful pioneering flights had been from west to east, taking advantage of the prevailing winds. And the princess was thinking not merely of an Atlantic crossing, from one coastal point to another; she was planning a capital-to-capital flight, something that would compare in achievement with Lindbergh's New York to Paris flight a month earlier. The cities she had chosen were London and Ottawa, two Empire capitals, accentuating the fact that this was to be a British project. The distance involved would be not much less than the Lindbergh flight, but success would be much more difficult because of the prevailing winds.

Both Minchin and Hamilton knew that several other people were laying plans for a similar flight. The princess was aware of it too. Most of the prestige would be lost if some other pair of fliers accomplished the crossing first. 'The American millionaire Levine is planning an east-west flight,' she told them. 'So are the French. All other British contenders seem to have dropped out. We were the first to fly the Atlantic from west to east and we've got to pull off this record as well.'

Adopting a more confidential tone, the princess lowered her voice. 'I've got a second ambition, too. It may go with the first, it may not. I don't know. We'll see how things work out. But I want to be the first woman to fly the Atlantic.' She turned to Dan Minchin, giving him her slight, impish smile. 'I want you to act as pilot, Colonel Minchin. Captain Hamilton told me he thought you might be interested. Will you do it?'

'I certainly will,' said Minchin. His ambition had for a long time been centred around an Atlantic flight and he had been planning one of his own in partnership with a pilot named Captain R. H. McIntosh, but the finance had fallen through. He had spent two years in Canada before the war, when one of his projects, interrupted by the outbreak of war, had been to form a civil airline company in Winnipeg. He had joined a Canadian regiment before transferring to the Royal Flying Corps, and the idea of a flight to Ottawa appealed to him greatly. 'It's just the chance I've been looking for,' he told the princess, 'if I can get leave, you can count on me.'

Thus Minchin was committed. But for Hamilton it was an entirely different matter. He had no ambitions to take on such a flight. Unlike Minchin he was a short-range pilot, with little liking for the tedium of endurance flying, and unaccustomed to long flights over the sea. He was happier flying the shorter overland hops in a light plane, running a luxury taxi service between the European capitals and playgrounds, as he had done that winter. But as the princess's personal pilot and friend he had promised to find her the right man. Now, lulled by the food, the wine and the music, and infected by Minchin's enthusiasm, he was suddenly vulnerable to the princess's charm.

'What about you, Captain Hamilton?' she asked. 'I expect Colonel Minchin will want a second pilot. Wouldn't you like to go?'

'Perhaps I would. Yes.'

He had said it. In an unguarded moment he had committed himself. The words could never be expunged from the record.

Hamilton's weakness was that he had created a popular image that was stronger than himself. Much as he regretted his decision next morning, he knew he could never back out.

Princess Loewenstein's wish that the attempt should be all-British was defeated by difficulties in getting a suitable British plane, and eventually she had to agree to the purchase of a Dutch Fokker monoplane – a single-engined Fokker VIIa – similar to the small passenger carriers operated by KLM. There was some consolation in the fact that the motor which they hoped would power them across the Atlantic was British. It was a 450-hp Bristol Jupiter radial engine, which gave the Fokker a cruising speed of about 110 miles an hour.

When Minchin and Hamilton got down to working out distances and fuel consumption they began to doubt whether a direct flight from London to Ottawa was possible. To make such a crossing, in the teeth of the prevailing winds, they would have to convert the plane into a flying petrol-tank. They would need a very long run to get the overloaded Fokker airborne, and no such field was available near London. In any case, with the ever-present risk of strong headwinds, it was surely more prudent to shorten the distance by starting from the west coast of Ireland.

On 26th July Minchin and Hamilton flew to Baldonnel airfield in Dublin *en route* for Clifden, on the far Galway coast, the remote spot where Alcock and Brown had landed after their Atlantic flight eight years earlier. But they were unable to find any suitable take-off run. The West of England seemed the next best bet, and Minchin suggested Upavon, on Salisbury Plain. There was a very long take-off run at Upavon, and the airfield was on a plateau, which might allow the plane to sink over the edge at the end of the take-off run without crashing. Also the place was sufficiently isolated to encourage the hope that publicity might be avoided. An added advantage was that they would be near the engineers at Bristol. Princess Lowenstein lobbied a friend at the Air Ministry, and permission to use Upavon was granted.

In the middle of August Minchin and Hamilton went to Amsterdam to collect the Fokker monoplane, and on 18th

August they flew it to Croydon and then on to Filton, where technical experts of the Bristol company gave the engine a thorough overhaul and retuning. Meanwhile the cabin seating was removed and eight additional fuel tanks were fitted, giving a total capacity of 800 gallons and an endurance of over 40 hours. From Upavon to Ottawa was approximately 3,600 miles. If they could average 100 miles an hour they would get there in 36 hours – exactly a day and a half in the air.

It required only the simplest arithmetic to show that a steady headwind of 20-25 miles an hour would put Ottawa out of reach. There would still be every chance of reaching Newfoundland. But winds of 50-60 miles an hour – and gales of this strength were not uncommon – could push even Newfoundland out of range. Thus much depended on choosing a period when favourable weather was forecast.

While the Fokker – named the *St Raphael*, after the patron saint of fliers – was at Filton, news came through of the progress of rivals. Two other recently-announced British attempts were said to be imminent, and the French and American planes were also reported to be nearly ready to go. Meanwhile the Fokker was still further delayed, Air Ministry officials having refused a certificate of airworthiness until modifications were made to the tail unit and rudder. The tail was similar in design to that of a Fokker VIIa which had recently crashed, and it wasn't until additional bracing wires had been fitted that the certificate was forthcoming. At last, on the night of Thursday, 25th August, work on the airframe was finished, and next day Minchin and Hamilton flew the machine to Upavon, moved into the Officers' Mess, and waited for favourable winds.

The hope that Upavon would be sufficiently remote to avoid publicity proved a naïve one. Reporters soon trailed across the Plain, creating all kinds of problems for Wing Commander Vernon Brown, the station commander, and Flight Lieutenant Freddie West, VC, his adjutant. Day after day the Air Ministry reported strong headwinds in the Atlantic, and Minchin and Hamilton – and the reporters – waited. Princess Loewenstein stayed in London, her intentions still uncertain. The strain of waiting, and the daily reports of the imminent departure of competitors, created an unbearable tension at Upavon. It was too much for Hamilton, whose nerve began to fail him. 'I don't think we'll make it,' he told Freddie West.[1] 'When I took it on I thought we had at least a 50-50 chance, but now I think we'll never get there.'

'Then why are you going?'

'I've just got to go through with it. I couldn't stand the ridicule if I backed out now.'

'What about the princess – is she going?'

'I don't think so,' said Hamilton. 'I think she's just in it for the publicity.'

While Hamilton remained gloomy but resigned, Minchin exuded earnest endeavour and confidence. 'You're not going to get there, are you?' asked Vernon Brown, putting into words the general opinion at Upavon. 'Oh, I think so,' said Minchin, 'it depends on the winds. If they're at all reasonable we'll get over. If they're not . . . well, someone'll make it sooner or later. I only hope it's an Englishman and not that fellow Levine.'

On Tuesday, 30th August, Hamilton had a phone call from Princess Loewentstein. 'Levine is ready to start,' she said. 'We've got to get away tomorrow. I've been on to the Air Ministry and the weather reports are quite favourable. I'm driving down to Upavon early in the morning and I'll be there at six-thirty. If everything's ready we'll go.'

So she really did intend to come. If Hamilton had ever felt that there might be some line of retreat for him, he knew it was gone now. 'I don't like it,' he told Freddie West. 'The weather isn't all that bad, but it's not as good as we'd like. We're being pushed into this by Levine.'

That evening in the Mess there was a farewell party for Minchin and Hamilton. Wives as well as officers crowded into the ante-room, pressed round the piano, and danced and sang. One tune that was asked for again and again was the hit song from a current West End show. The song was called *My Heart Stood Still.*

[1] *Winged Diplomat,* the Life Story of Air Commodore Freddie West, VC, by P. R. Reid (Chatto and Windus).

I took one look
At you,
That's all I had
To do,
And then my heart stood still....

For Hamilton the song had a special fascination. It brought back many memories. He had entertained the whole company from the show at his flat. The tune had been played that night at the Savoy. It was at his request that the pianist repeatedly returned to this haunting, wistful melody.

By eleven o'clock the ante-room began to empty. Everyone wanted to be up early next morning to see the plane off. To ensure a daylight landing in Ottawa the *St Raphael* had to be airborne by seven-thirty. At midnight, when Minchin, quiet and optimistic as ever, went to bed, there were only three or four people left. But Hamilton still sat near the piano, a glass in his hand, a cigarette between his lips. In vain he was urged to snatch some rest. 'There'll be plenty of time to sleep when we get there,' he said, 'and a permanent rest if we don't.'

At five o'clock next morning the engine of the Fokker monoplane was carefully warmed up. The ten fuel tanks had already been filled to the brim. Roy Fedden (now Sir Roy Fedden, then chief engineer of the Bristol Aviation Company) and two of his senior staff, Frank Able and Freddy Mayer, were in charge. Minchin, his teeth gleaming white against his dark moustache and swarthy skin, was there. So was Hamilton, pale and drawn, his eyes ringed and bloodshot, struggling to shake off a hangover.

The normal wind direction at Upavon coincided with the longest stretch of grass, and the Bristol engineers had positioned the plane on the edge of the field to get the best take-off run. Now, however, the abnormal wind direction that was making the flight feasible threatened to make the take-off impossible. But on the far side of the Upavon road there lay an unused section of the aerodrome where it might be possible to get a fair take-off run into wind. There was one drawback – the field sloped downhill and was bordered by the hollow through which ran the Upavon road.

This unused stretch of aerodrome nevertheless offered the only prospect of a successful take-off that morning, and Minchin paced out the distance with Freddy Mayer. Minchin thought he could make it, but Mayer was all for caution; he arranged to station an ambulance at what he considered the safe limit, the point by which Minchin must be airborne. If he wasn't safely off by then he must cut his engine. Minchin nodded his head.

It was a dull, misty morning with an overcast sky, but at 6.30, when Princess Loewenstein arrived in her chauffeur-driven Rolls, the sun broke through fleetingly. For a moment, as she emerged from the car outside the Mess, some of the officers peeping through the window of the ante-room felt a compelling urge to laugh. She was dressed in purple riding-breeches, over which she had drawn a pair of fur-lined yellow top-boots and a belted, purple leather jacket, the whole set off by a black toque hat. Her chauffeur was unloading her luggage, which included a wicker armchair, a hamper, and two hat-boxes – an incongruous mixture. But her petite charm and simple, unaffected manner soon disarmed everyone.

'Thank you for looking after my pilots,' she said as she shook hands with Vernon Brown, 'and how very kind of you and your brother officers to be up at this unearthly hour to meet me.'

'We've got some coffee for you,' said Vernon Brown.

'That would be wonderful.'

'How do you feel?'

'Fine! But – could I have a glass of brandy with my coffee?' It was the only hint she gave of the nervous tension she must have been feeling.

'Do you really mean to go with them?' asked Freddie West.

'Well, I'm dressed for it!'

As they talked, a second car drew up outside the Mess, and from it stepped the Roman Catholic Archbishop of Cardiff, Archbishop Mostyn, complete with flowing white lace robe and biretta. Two other priests were with him. The princess had asked the archbishop to bless the plane before take-off.

'I hope he won't smell that I've just had a brandy,' whispered the princess, her face bright with mischief.

Minchin was beginning to get restive. 'We mustn't waste any more time,' he was saying. 'We must get away now.' Soon the whole party moved off down to the airfield, the archbishop travelling in the Rolls with the princess. Meanwhile a group of RAF mechanics had wheeled the *St Raphael* from the main airfield to the unused section. The sky was overcast again, and skeins of misty rain were drifting across the Plain. Light south-west winds were forecast in mid-Atlantic, but a following wind was predicted up to about half-way.

Minchin, in a dark lounge suit and double-breasted waistcoat, white shirt and collar and a soft felt hat, looked fresh and fit. He might have been going for a morning stroll. 'Everything is in our favour,' he said. 'The wind is right, the weather's good, the plane's in perfect trim. But don't ask me how long we shall take to get there. It may be 20 hours, it may be 40.'

'But you're sure of getting there?'

'Of course. Do you think I would go otherwise?'

The princess too spoke confidently. 'But whatever happens,' she told reporters, 'I'm proud of being the first of my sex to make the attempt.'

Hamilton, deathly white now but clearer of eye (he had been violently sick behind one of the hangars) failed to hide his fears. 'This is a grim ordeal,' he kept saying, 'a grim ordeal.' His, perhaps, was the highest courage of all, not the courage of a man who does not know or feel danger, but that of a man who forces himself against every natural instinct to do what is expected of him. Already he had told a friend that he would make a good meal for the sharks. Now he pressed a wad of £1 notes into Freddie West's reluctant hand. 'Here's £25, Freddie,' he said. 'All my spare cash. Will you give it to the RAF mechanics who've been working on the plane? It's better they have it than the fishes.'

The wreck of the *Lady Be Good*

SUNDAY, APR. 4, 1943

Naples — 28 planes — things pretty well mixed up — got lost returning, out of gas, jumped, landed in desert at 2:00 in morning, no one badly hurt, can't find John, all others present.

MONDAY 5

Start walking N.W., still no John. a few rations, 1/2 canteen of water, 1 cap full per day. Sun fairly warm good breeze from N.W. Nite very cold, no sleep. Rested & walked.

TUESDAY 6

Rested at 11:30, sun very warm no breeze, spent P.M. in hell, no planes, etc. rested until 5:00 P.M. walked & rested all nite, 15 min. on, 5 off.

WEDNESDAY, APR. 7, 1943

Same routine, every one getting weak, can't get very far; prayers all the time, again P.M. very warm, hell. Can't sleep every one sore from ground.

THURSDAY 8

Hit Sand Dunes, very miserable, good wind but continuous blowing of sand, every one now very weak, thought Sam & Moore were all done. La Motte eyes are gone, every one else eyes are bad. Still going N.W.

FRIDAY 9

Shelly, Rip, Moore separate & try to go for help, rest of us all very weak, eyes bad, not any travel, still very little water. nites are about 35°, good N. wind, no shelter, 1 parachute left.

A page from the diary of Lt Robert F. Toner

The take-off of the *St Raphael*

Leslie Hamilton and F. F. ('Dan') Minchin

The Tudor *Star Tiger*

Douglas Corrigan (*left*) on arrival at St Louis, August, 1938

Elsie Mackay and Ray Hinchliffe at snowbound Cranwell just before take-off

At this moment the archbishop strode through the drizzle to the plane. Sheltering under the high wing, with a priest on either side of him and with Princess Loewenstein in the rear, he sprinkled holy water over the machine. Then he turned to the princess. 'God bless you all,' he said, 'may you have a safe journey. We will not forget to pray for you.' The princess sank to her knees to receive the archbishop's benediction, kissing his hand.

Minchin and Hamilton now joined the princess. The archbishop grasped each man firmly by the shoulders and looked into his eyes, wishing him good luck and God Speed. Hamilton's demeanour no longer betrayed his fear. The strain of waiting was over. He turned steadily after Minchin and followed him into the plane.

Princess Loewenstein clambered in after them, helped by the archbishop. Then she crawled forward over the petrol tanks to her wicker chair immediately aft of the cockpit, where her luggage and the two hat-boxes were already stowed. There she sat with her head resting against 600 gallons of petrol.

One of the mechanics reached up at the propeller, turning it slowly. Minchin clicked on his switches, the mechanic dragged the propeller sharply downwards, and the engine exploded into life. 'Let's go,' said Minchin.

Many people still doubted whether the heavily-loaded *St Raphael* would get off; they expected it to crash into the sunken road which bordered the field, or into the rising bank and stake-and-wire fence on the far side. If that happened, the blaze ought to be something to see. Vernon Brown and Freddie West jumped into their cars and were about to set off with the ambulance and fire-tender towards the bottom of the sloping field when they were stopped by Roy Fedden.

'Where are you going, V.B.?'

'Jenner's Firs. Right at the end of the field. That's where they're going to hit the bank.'

'I'm going to a point about fifty yards further on. That's where *I* think they're going to hit the bank.'

Three minutes later, shortly after half-past seven, the

Fokker monoplane began to roll forward heavily, like a bloated animal, lurching sloppily across the carpet of grass. The rain had stopped, but the mist was settling, and the plane soon disappeared from view over the brow and down the slope. Several reporters bumped across the airfield in their cars in pursuit.

Down at Jenner's Firs the high wing, widely-spaced undercarriage and the radial snout of the Fokker suddenly appeared at the top of the slope. The take-off run had still been something of a compromise between length and wind direction: Minchin was using just about the longest stretch of grass on the unused section, but it was not quite dead into wind. He reached the point where Mayer had stationed the ambulance, and passed it, and still the plane seemed firmly anchored to the ground. Yet he did not throttle back. Fedden, Brown and West, pilots themselves, fought with imaginary controls as they willed the plane safely away. At 200 yards distance the tyres were still bulging with the plane's weight. At 100 yards the plane, apparently flat out, still gave no sign of lifting.

The *St Raphael* was racing for the road and the bank on the far side with what seemed an irresistible self-destructive urge, right between the vantage points chosen by Fedden and Brown. Freddie West shouted an alarm to the men in the fire tender. 'Look out! Stand by! It looks like trouble!'

The Fokker was hurtling forward at full power, its engine roaring a strident warning to the men at the end of the field. Seventy yards, fifty yards, and still it had not come off. 'Lift!' shouted West. 'For God's sake lift!' Every muscle in his body was straining to push the plane upwards. He could see Minchin in the cockpit, staring straight ahead, apparently unmoved at his peril. Then in the next instant, as though swept by machine-gun fire, everyone at Jenner's Firs dropped flat on his face.

Like a vast bat of destruction the Fokker careered down to the edge of the road, on the far side of which it must crash and disintegrate. It was at that moment that Dan Minchin yanked the stick back into his stomach. The plane staggered up momentarily, lifting no more than five or six feet, but it was just enough to clear the road-bank and the barbed wire fence. Freddie West, still flattened on the ground but now looking up, could not restrain a hoarse shout of relief.

'My God, Minchin! Well done, Minchin! You certainly knew how to cut things fine.'

Indeed nothing could have been more finely judged. The airmen at Jenner's Firs felt all the elation of men who have themselves escaped death. And now, as the plane sank back towards the ground after Minchin's rough usage, the downward slope at the edge of the plateau saved it from crashing. All this Minchin must have calculated before take-off.

Slowly the plane settled down to a steady course before disappearing into the mist. The first Atlantic attempt by a woman had begun. Twenty-five minutes later the *St Raphael* flew low over Filton, exactly on course. The heavy overload was preventing Minchin from climbing, but the plane was maintaining height.

The latest weather reports, received shortly after Minchin's take-off, disclosed that the following wind which had earlier been forecast for the first part of the flight would rapidly disappear. They could now expect a slightly adverse wind as far as the western Irish coast, which would stiffen in mid-Atlantic. Unfortunately no message could be got to Minchin because he carried no radio, but he would have plenty of time to estimate the wind by reference to his computed speed while flying over land before committing himself to the Atlantic crossing.

Minchin followed a course which took him along the uneven South Wales coast, sometimes over the land, sometimes over the sea. At 9.20, nearly two hours after take-off, he passed clear of the Pembroke coast and out into the Irish Sea. At ten o'clock he made a landfall south of Wexford, and his plane was seen to cross the coast at 500 feet.

It had taken the *St Raphael* two and a half hours to travel 200 miles – an average of only 80 miles an hour. Their airspeed would improve as they burned off fuel, but this would be offset by increases in wind strength. Minchin decided to make another check when he reached the western Irish coast. His airspeed should have improved by then.

Thurles, in Tipperary, was their next pinpoint. Already Minchin was finding the controls less sluggish and the plane more airworthy, enabling him to reach just under 1,000 feet. His next – and last – check would come at the most northerly point of the Aran Isles, at the entrance to Galway Bay. By the time they got there the *St Raphael* had settled down at 1,000 feet and was handling well. The groundspeed had gone up to 90 miles an hour, just as Minchin had expected. As they used up more and more fuel it would improve still further, giving an ample insurance against increased winds. They might not get to Ottawa, but their margin for the Atlantic crossing was satisfactory. Minchin had no doubt that barring accidents or unforecast gales it was enough. By far the most dangerous part of the venture had been the take-off.

Hour by hour through 31st August the *St Raphael* flew steadily on course, progressing far out into the Atlantic. There were no storms in her path. There was a risk of icing, but the headwinds were steady at 30 knots. There was every reason for optimism. Hamilton's wife, who was still in New York, left by train for Ottawa when she heard news of the take-off. A reconciliation looked probable. The Prince of Wales and Prince George, on an official visit to Canada and due to leave Ottawa that day for Montreal, announced that they would postpone their departure if the plane was known to be approaching Ottawa.

When darkness overtook the *St Raphael* and the grey waters of the Atlantic merged with the black horizon, the cockpit lights glowed reassuringly. The Jupiter engine purred and vibrated comfortingly, the revolutions steady, the petrol and oil consumption normal. The Fokker monoplane was nearly half-way across the ocean, ploughing through the long Atlantic night.

Soon after half-past nine that evening, after fourteen hours in the air, nine of which had been spent out of sight of land, the *St Raphael* was sighted for the first time by a passing ship. It was the oil tanker *Josiah Macy*, bound for Germany. Spotting the plane's navigation lights overhead at about 1,000 feet, the crew signalled with an Aldis lamp, and the *St Raphael* replied with a succession of flashes from its own signalling lamp in recognition and reassurance. Soon the radio operator of the tanker was transmitting to the world the news that the *St Raphael* was in mid-Atlantic, holding her course for the direct flight to Ottawa, her engine running smoothly and strongly, her crew alert and confident. Since leaving the Irish coast the plane had covered nearly 900 miles against head-winds which had reduced her average speed to about 85 miles an hour. But this speed could still be expected to improve as more fuel was burned and the plane's load lightened. The margin for Ottawa was still just adequate.

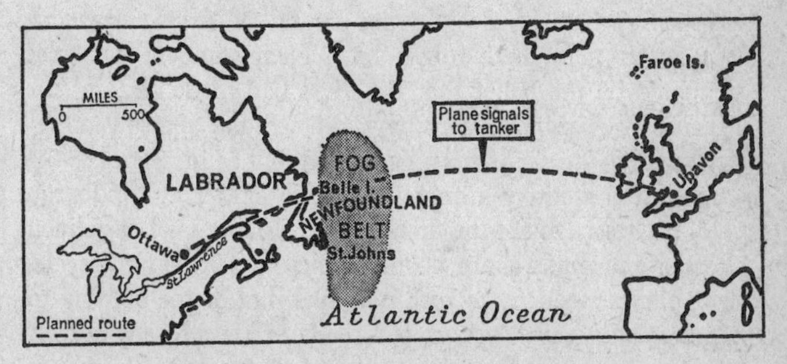

3. Upavon to Ottawa

With this definite confirmation of the flight's progress, preparations to greet the first successful Atlantic fliers from east to west went forward amid great excitement in both Newfoundland and Canada. In Ottawa, where there was to be a civic welcome, powerful searchlights had been installed and rockets were ready to guide the fliers in case they arrived after dark. At Harbour Grace, in southern Newfoundland, where Minchin planned to land if he felt he could not reach Ottawa non-stop, a line of oak barrels had been saturated with oil so that the airfield could be kept floodlit all night. The powerful Cape Race radio station, in the south, kept in touch with ocean liners for news of further sightings. And all lighthouses were asked to keep a careful look-out, especially in the Belle Isle region in the extreme north, which lay on the plane's direct route to Ottawa. Telegraph offices along the coast were kept open, and although the plane was not expected until dawn, a large force of volunteer watchers kept an all-night vigil, sending up flares at intervals. It seemed that the plane could not slip by unseen.

But during the night the vision of the watchers was impaired by banks of fog which drifted in off the ocean. It was the treacherous, opaque Newfoundland fog, which stretched for nearly a third of the way across the Atlantic, and through which the *St Raphael* would have to fly. Flares were sent up throughout the morning, and towards midday the fog cleared, but there was neither sight nor sound of the *St Raphael*. Speculation that there might have been some mishap was attenuated by the belief that the fliers had made straight for Ottawa. They might easily have crossed the sparsely-populated north in the early-morning fog without being seen.

Soon after midday a crowd began to gather at Lindbergh Field in Ottawa to see the fliers arrive. Hundreds of cars lined the field and many more people came on foot or by bicycle. Light planes stood by to rush newsreel film of the landing to Montreal and New York. But the lack of news of any sort of sighting since mid-Atlantic made people uneasy. Slowly, as dusk deepened into darkness and the evening wore on, uneasiness gave way to anxiety, until at last the crowds began to disperse.

As always in these cases, there were many reports that the *St Raphael* had been seen. All proved on investigation to be false. Next day, Friday, 2nd September, the belief crystallized that the princess and her two pilots had failed to complete the crossing. The odds, it was felt, had been against them. Fog off Newfoundland, coupled with the strong south-westerlies, must have defeated them when they were almost within sight of their goal. It was yet another story of a gallant attempt that had failed.

Over the years writers have painted a vivid picture of the last hours of the *St Raphael*, as its crew struggled on through the swirling Atlantic fog, finally running out of fuel and descending with dead engine into that cold, spumy sea. This was the fate of so many of the Atlantic pioneers. Yet it may be instructive to look at the facts again.

The Fokker VIIa was a well-tried machine, not likely to suffer from structural failure. A suspected weakness in the tail unit had been specially strengthened under Air Ministry guidance shortly before the flight. The Bristol Jupiter engine had been tuned up to concert pitch by the company's experts and had run sweetly for at least 14 hours – why should it suddenly fail?

At the time of the sighting by the *Josiah Macy* the plane was making good progress and seemed to have a fair chance of reaching Ottawa. Certainly it had an ample margin for Newfoundland, assuming the headwinds did not stiffen into a gale. As no such gales were reported, the *St Raphael* could hardly have run out of fuel before reaching Newfoundland. Icing, too, can similarly be eliminated; there were no further reports of icing conditions at the height at which the plane was flying.

What, then, can be the answer?

The south-westerly wind, as well as impeding progress, would very likely have blown the plane north of its course, especially in those last few hours of fog when measurement of drift was impossible. It could have been, therefore, that the *St Raphael* made her landfall in thick fog to the north of Belle Isle – somewhere, in fact, in the Labrador. Even when they finally emerged from the fog, Minchin and Hamilton might find themselves irretrievably lost. In that event, sooner or later the plane would come down in one of the remotest, most sterile, inhospitable voids in the world. It would then be a thousand to one against the crew surviving to reach civilization. Even the most intensive search might not find them.

Just as the arid wastes of the Sahara hid the fate of the Liberator *Lady Be Good*, so the icy snows of the Labrador may have frozen around their secret. Thirty-five years or more after the event it could still be that Dan Minchin and Leslie

Hamilton will one day be established as the first men to fly east to west across the Atlantic, and that the fulfilment of Princess Loewenstein-Wertheim's ambition to be the first woman to make the air crossing will be posthumously but sensationally revealed.

4

THE WORLD OF ALBERT VOSS

THAT passenger who sits next to you in the jet airliner as you sweep across Europe or over the Atlantic, the one who seems glumly engrossed in his thoughts and who says and does nothing except to issue a curt order to the air hostess for a Scotch and soda – do you ever wonder about his background? The question could be a pertinent one, because his state of mind may be important to you. It could even be a matter of life and death.

Of the twelve passengers who left Brussels for Croydon on the Imperial Airways Argosy airliner *City of Liverpool* at 12.36 on the afternoon of 28th March 1933, none had a more intriguing background than the man who sat alone in the rear seat on the starboard side opposite the luggage rack – Seat No. 20. His name was Albert Voss. Sixty-eight years old, and a German Jew by birth, he was a naturalized British subject who for many years had practised as a dentist in Manchester. Spruce and well-preserved, with thick well-brushed hair that was only lightly streaked with grey, he wore a heavy moustache that tended to droop at the edges giving him a look of melancholy power. Dark suiting and a butterfly collar lent him an Edwardian elegance, and he looked at least ten years younger than his age.

In the past few years Voss had taken to augmenting his income as a dentist by trading in dental appliances, and his frequent visits to the Continent – he was reputed to go to Germany five or six times a year – were ostensibly to buy dental equipment, including teeth, for resale in England. There were rumours, though, of other less reputable deals.

A man of strong physique and boundless energy, Voss handled substantial sums of money and lived extravagantly, yet over a period of many years he was always in debt. He was an undischarged bankrupt. No one seemed to know what happened to his money. Of his five children, two sons were practising dentists, but there was some degree of estrangement with the family as a whole through a second marriage two and a half years earlier to a woman 25 years his junior. He displayed wide extremes of temperament, ranging from unusual coolness to high excitability, and during the excitable periods he was subject to a kind of paroxysm which sometimes lasted several minutes. More recently he had suffered from headaches and fits of depression, and once, after failing to get a loan from one of his sons, he had talked of suicide. 'I may as well blow my brains out,' he had said. But no one had taken this seriously.

Four months before this trip to Brussels he had collapsed after taking an overdose of aspirin. About this time, Scotland Yard were said to be enquiring into his business dealings on the Continent. There was talk of drug trafficking and smuggling, but no charges were brought. Then, announcing that his trading in Germany was threatened by the advent of Hitler – again there were rumours, this time of fraud charges awaiting him in Berlin – Voss had arranged this trip to Brussels. With him, travelling on the same plane, was a business associate named Dearden, a much younger man, but a man of doubtful reputation believed to be concerned in smuggling activities.

At eleven o'clock on the morning of the return flight from Brussels, Voss and Dearden had taken a taxi from their hotel to the airport. There they had taken out 35-franc insurance policies giving a benefit of £500 to their dependants in the event of death through air accident in the next 24 hours.

The *City of Liverpool*, which for five years had plied regularly on the route between London, Brussels and Cologne, had left Cologne that morning on the return flight to London at 10.22 – 22 minutes late. The captain, the 35-year-old Lionel Leleu, had been with Imperial Airways since 1926, following service with other air companies and with the RAF. He had over 4,000 hours as first pilot, and an unblemished record of civil flying. Leaving his home in Purley the previous morning to motor to nearby Croydon Airport, he had promised to take his wife out for a treat that evening if he got back to schedule. She was expecting their third child.

Only four of the eighteen available seats – the two rear seats on the port side had been removed to provide space for the luggage rack – were filled when the plane left Cologne. One of these four passengers, the 16-year-old Fraulein Lotte Voss, was on her way to finishing school at Wimbledon. What her relationship was with Albert Voss was later the subject of comment.

The airliner made good time travelling due west from Cologne and was almost back on schedule when it landed at Brussels. It was due to leave for Croydon at midday, but there was some delay for refuelling and the loading and unloading of freight. Eight passengers got on at Brussels, making twelve in all. By the time the plane had been refuelled and the freight and baggage redistributed – most of the passengers' luggage was housed in the rack at the rear of the cabin – it was well after midday. When the plane eventually took off it was 12.36.

The route lay first across the flat plains of Belgium south of Ghent to the coast. Then came a 50-mile Channel crossing, followed by a pleasant flight across the orchards of Kent to Croydon. The estimated flight time, at 95 miles an hour, was exactly two hours. It was a bright, crisp spring afternoon, with hardly a breath of wind, and the smoke from the chimneys of the Belgian villages rose almost vertically. The sky was cloudless, and visibility was limited more by sunshine than by haze. Through the wide cabin windows, stretching almost the whole length of the fuselage, the passengers on the starboard side could pick out an express train gliding along the Brussels-Ghent-Ostend railway four thousand feet below. Some of the train's passengers, perhaps, had a Channel crossing by sea ahead of them. They wouldn't get to London until next day. Captain Leleu, though, in spite of the minor frustration of being 36 minutes behind schedule, would be home in Purley in time for tea.

The passengers on this first of the luxury airliners relaxed comfortably in wicker armchairs. There were two blocks of these chairs, set in pairs on either side of the fuselage, with a gangway down the middle. As an experienced air traveller, Albert Voss had chosen what by general consent was the safest seat in the plane – the rear seat, nearest the door.

The life of an airline pilot in 1933 was splendidly uncomplicated. He took off, he flew the plane from airport to airport, and he landed. The problems and responsibilities of flight and navigation were much more personal and urgent than they are today, and all the decisions were his. But there were no complicated traffic patterns, no sophisticated control procedures, no cumbersome let-downs. The sky was virtually empty. The problems exercising the minds of the passengers, though, were much the same as they are today. All had their worries – more fundamental in some cases, perhaps, than such trifles as getting home in time for tea.

Twenty-four minutes after take-off, at one o'clock, the *City of Liverpool* passed due south of Ghent. The wireless operator scribbled his half-hourly entry in the instrument log. 'All engines running normally,' he wrote. 'All instruments reading correctly. Height 4,300 feet.' The Argosy was on the homeward leg of yet another round trip and all was well. Albert Voss was still sitting in Seat No. 20, several rows behind his associate Dearden, and some distance away from Fraulein Voss. Another 20 minutes and the wireless operator had tuned in his transmitter to the home frequency and called Croydon. 'We are now passing Roulers,' he signalled. 'Time, 13.20.' In 15 minutes they would be over the water. It promised to be a tranquil crossing. But five minutes later, at 13.25, something went wrong in the *City of Liverpool,* something which has never been explained to this day.

There were many eye-witnesses of the subsequent flight and crash of the Argosy, but none could penetrate into the cockpit or cabin during its slow, controlled plunge earthwards. Still less could they penetrate into the minds of passengers and crew. Several Belgian peasants had seen the plane fly over, slowly but easily pursuing its course. One, a farm labourer, working in his allotment garden, thought he saw a wisp of smoke from the top of the aft end of the hull. This was followed by a tiny spout of flame near the tail. Then, in a purposeful but shallow dive, the plane began to lose height.

Almost before the plane had begun its dive, a single black object fell away from the fuselage and dropped earthwards. It looked like a sack, or some loose item of luggage. In fact it was one of the passengers, unharmed except for very slight singeing of the moustache and eyebrows. The man appeared to have no parachute. It was Albert Voss.

As the outline of the plane sharpened, a bright blue flame was visible through the windows at the back of the cabin. The fire had taken hold with such ferocity that no one in the plane had had time to operate the hand fire-extinguishers carried for just such an emergency. The only hope, it seemed, was to get the plane on the ground as soon as possible. Captain Leleu, holding the machine on an even keel, continued his controlled descent, not too steep but as rapid as the plane's structure would safely allow.

There was no fire in the front of the aircraft, no damage to the three engines and none to the wings. Pilot and wireless operator were unhurt. In the cabin, only one passenger, a woman seated immediately in front of where Voss had been sitting, had suffered any burns. There was still every chance that Captain Leleu would get the plane down in time to save the lives of all but one of his complement.

The plane sank over the heads of eye-witnesses and carried on towards a line of poplars fringing the road into Eesen, two miles south of Dixmude. Now it was no more than 300 feet up. Beyond the poplars was a large open field at which Captain Leleu was evidently aiming. In perhaps a quarter of a minute the wheels of the Argosy – it was a fixed undercarriage – would touch the ground. If Leleu could make a successful landing, exit from the front of the plane would be safe enough.

At 250 feet the *City of Liverpool*, its name now easily readable from the ground, was still sinking rapidly and steadily. Lionel Leleu could be seen standing up in the cockpit, desperately trying to get some response from the controls. Then there were two echo-less, staccato sounds, like pistol shots, and somewhere near the cabin door the Argosy split in half. A ball of released flame masked for a moment the point of separation, and then like a severed worm the two parts curled away.

The front section, intact and undamaged, instantly turned over on its back and dived into the ground with great force. Captain Leleu had turned off the petrol cocks, but the fuel tanks burst on impact. The whole front portion, hitherto untouched, became an inferno of flame. Meanwhile the aft portion fell separately, while fragments of the tail section splintered and fell some distance away. A woman passenger – the one who had been sitting immediately in front of Voss – was thrown out as the plane broke up. As she fell, her clothes billowed out in the shape of a parachute. Around her, baggage from the aft freight compartment was scattered and strewn over hundreds of yards.

All the people still on board when the two separate halves of the airliner crashed were killed instantly. There were no survivors. It was the worst disaster in the 14-year history of British civil aviation, and the first fatal accident involving an Argosy in six years of operating. What had gone wrong? British and Belgian accident investigation teams moved at once to the scene of the crash, and Imperial Airways carried out their own independent enquiry. It was obvious from eyewitness accounts and from a study of the wreckage that the initial cause of the accident had been fire. But what had set it off, and where had it started?

In spite of the violent fires that burned in the front portion of the plane after the bursting of the fuel tanks, the investigators were able to prove that there had been no trace of fire in this half of the aircraft prior to the crash. Freight and components from the forward compartment that had been buried in the ground at the time of the crash were untouched by fire. Fire in the engines or in the fuel and exhaust systems was similarly ruled out. Eventually the source of the fire was narrowed down to a tiny area at the rear of the passenger cabin, either in the lavatory or the luggage rack. Wreckage from the rest of the plane, both front and rear portions, was recovered in recognizable state, but nothing identifiable was found from the lavatory, and of the whole of the baggage in the luggage rack, and of the rack itself, there was absolutely no trace.

Of the 18 pieces of passenger luggage recorded on the manifests, twelve had been put in the luggage rack, including all the baggage put on the plane at Brussels and a tin trunk belonging to Lotte Voss. Somewhere amongst this luggage, or possibly in the lavatory, the fire had started. It seemed to the investigators that there were two main possibilities. First was the deliberate firing by time-fuse of an incendiary type of bomb, or of some other inflammable material, either in the lavatory or the luggage rack. Such firing could be a simple case of sabotage, or it could have been done by a passenger, although this seemed unlikely. Second was the accidental or spontaneous firing of some combustible material, perhaps through carelessness with a cigarette or through chemical action or vibration. All other possibilities were eliminated.

Two days after the crash the bodies of the victims were flown home and funeral arrangements made. Albert Voss was to be buried in the Jewish cemetery outside Manchester at 3.30 on the Sunday afternoon following the tragedy, and his body was brought by road from Croydon, reaching Manchester on Saturday evening at five o'clock. But meanwhile, the circumstances under which Voss had left the aircraft, over a mile from the subsequent crash, and before the fire had gained a hold, judging from his very minor burns, had attracted the suspicions of all the investigators. On that Saturday morning, Imperial Airways had called in Scotland Yard, and fifteen minutes after the body reached Manchester two CID men arrived at the undertakers and removed the coffin. The body of Albert Voss was to be subjected to a searching analysis; meanwhile the funeral was indefinitely postponed.

What did the pathologists expect to find? All the circumstantial evidence pointed to some connection between Albert Voss and the fire. He had been the nearest to both lavatory and luggage rack. He had been sitting alone, where he could not be easily observed. He had evidently been the first to discover or to be affected by the fire. He had jumped instantly, virtually before he was hurt. As a man who bought and sold anaesthetics he had access to highly inflammable substances. He was said to be deeply in debt. He was a suspected smuggler and drug trafficker – perhaps a drug addict. He had spoken of suicide and had possible attempted it once already. But even more appealing was the suspicion that he had attempted to disappear. Rouse and the car murder of the previous year were fresh in everyone's mind, and it seemed a reasonable deduction that Voss had intended to set fire to the airliner and then make good his escape. Amongst the charred and scattered wreckage the absence of one passenger might be impossible to detect.

All this would have made sense if there had been any evidence that Voss had carried a parachute. But there wasn't. He had jumped without one. At least, no trace of a parachute could be found, and the wreck and the countryside around it were combed for one. Was it possible that the parachute had caught fire at the time of the act of arson? All the evidence was that the fire had ignited rapidly and fiercely and that anyone sitting in Seat No. 20 must have been in immediate danger. Did Voss concoct this diabolical plan, and did it founder at the last moment in this way?

It seemed incredible that any sane person could contemplate wholesale murder of innocent people with so dubious a chance of preserving his own life and deceiving the investigators. Could Voss have been insane? Was he, perhaps, under the influence of drugs? This at least an autopsy would show.

'Let me try to fathom,' said the coroner at the inquest, addressing one of Voss's sons, 'these grave, suspicious circumstances in which your father met his death.' Then the little world of Albert Voss was ruthlessly exposed. Was it not odd that a man in his impecunious state should travel to Germany by air six or seven times a year? What was it that Scotland Yard had been probing four months earlier? What stability could one expect from a man who had talked of suicide and once apparently attempted it? And who was Lotte Voss? Was she a relation? None of the family had ever heard of her, but they admitted that there might be a number of nieces in Germany whose names they would not know.

'Would you be surprised to hear,' asked the coroner, addressing another of Voss's sons, 'that she wasn't a niece at all?'

'No.'

The inference was obvious. In addition to being a swindler, a smuggler, a drug trafficker, and a suicide, Albert Voss was a womanizer. Calumny and opprobrium could go no further. It was a short step from here to wholesale murderer. In vain did the family insist that Voss, although careless over money matters and unfortunate sometimes in his business associates, was a practising dentist and a genuine trader in dental equipment, was no smuggler or drug trafficker, still less drug addict, was a good husband and father, and was certainly not a suicide. None of the terrible insinuations about Voss had been proved, but he had been given a bad name. He was about to be metaphorically hung.

The inquest was adjourned for the pathologist's report. When it came it was disappointing. There was no evidence of drugs, no trace of poison, nothing to throw any further light on the circumstances of the exit from the plane. Albert Voss had died of multiple injuries resulting from his fall. Theorists were back with the hypothesis of the missing parachute. Others who could no longer believe that Voss had deliberately set fire to the plane still blamed him for the disaster. He had been smoking – and smoking was prohibited. Or he had been smuggling some highly inflammable chemical in his luggage. Hecolite, a solvent used in the making and repairing of false teeth, was mentioned by the experts. It was highly inflammable in its liquid state. So Voss had been carrying hecolite. Again there was no evidence to support any of these theories. Voss very seldom smoked. And every scrap of wreckage had been tested for traces of an inflammable chemical without success.

When the inquest was resumed, and after the pathologist's evidence, the coroner addressed the jury. 'From the available evidence,' he asked them, 'can you possibly come to the conclusion that Voss was in any way responsible for the liner catching fire by accidental means or with the deliberate intention of taking his life and in doing so the lives of fourteen others?

'It seems incredible, unless Voss lost his reason. You might think that the case is not entirely free from suspicion, but suspicion is not enough. In the absence of definite and direct evidence you would not be justified in saying that Voss was guilty of such a terrible crime.

'You will probably think that there is no evidence on which you can possibly return a verdict of suicide, and not sufficient evidence to show that he was in any way responsible for the fire.'

But the jury, their minds still loaded with suspicion, rejected what amounted to a direction from the coroner to reach a verdict of Accidental Death. After an hour's discussion they returned an open verdict.

A spokesman for Imperial Airways was present at the court. 'In fairness to the relatives,' he said, 'I say that Imperial Airways do not, and have not, suggested that Mr Voss deliberately set fire to the *City of Liverpool*.' But the open verdict had left a stigma on the name of Albert Voss which was never removed.

The persecution of Voss's family, which had begun before the inquest, went on. Cheated of the vilification of Voss that they had expected from the coroner, a section of the public vented its spite on his unfortunate widow. Three months later, sickened by slander, sneers and scurrilous letters, she disappeared; her body was later found in the Manchester Ship Canal.

At this distance from the tragedy it seems clear that the clue to Voss's guilt or innocence lay in his state of mind when he embarked on the flight. A desperate man, particularly a man subject to brainstorms, as Voss apparently was, might be capable of anything. Was there any available evidence which might have thrown further light on the state of mind of Albert Voss on the day of the disaster? Strangely enough, it seems that there was quite a lot.

What did Voss and his partner go to Brussels for? This question is surely fundamental. There is evidence to show that they went there to do, and that in fact they did, perfectly legitimate business. They had flown to Brussels four days earlier, on 24th March. After enquiring if there was a casino in Brussels, and being told there was not, they moved out of their hotel near the main station within an hour of arriving and spent the next two days gambling at the casino at Spa. Here perhaps is the clue to what happened to Voss's money over the years. Whether they made money or lost it on this occasion is uncertain, but what evidence there is – their demeanour and their readiness to spend money – suggests that they may have won. On 27th March they returned to Brussels by train and booked in for two nights at their original hotel. They were due to return to England two days later, on the 29th, and their air bookings were already made.

On the same day, the 27th, they visited the managing director of a firm of dental appliance manufacturers and ordered goods, including teeth, to the value of £1,250. M. Bogaert, the Belgian with whom they dealt, was impressed with Voss as a shrewd business man who drove a hard bargain. It was arranged that they should come back to settle details of packing and delivery next day.

Soon after nine o'clock next morning, Voss and Dearden were back in M. Bogaert's office. This time they were in a hurry. They had had, they said, a cable from England. 'We've been called home unexpectedly,' said Voss, 'but we'll be back again soon for a longer visit.' He then gave M. Bogaert a cheque for £400 on account. The two men had already contacted the airport from their hotel and brought their air bookings forward 24 hours.

It could be that the key to Voss's state of mind lies in the mysterious cable. But this would not appear to be so. Whatever the message was, it did not alter his business plans in Belgium. The solid evidence is that Voss had made a new contact in what for him was a new country and that he was pleased with it. He had managed the deal in a business-like manner, and he intended to come back and follow it up. Something urgent was awaiting his attention in England, but nothing that threatened to disrupt his future plans in Belgium.

Could the cable have been from or about Lotte Voss, to say perhaps that she was travelling that day, and asking Voss to meet her and travel with her? There is nothing to support such a theory, no evidence that Voss ever spoke to her or even knew of her existence. Voss is a very common German name, and in all probability they were completely unconnected. It is known that Lotte Voss was being met at Croydon by the headmistress of her finishing school. The suggestion that she was travelling with Voss, and the sneer about their relationship, seems to have been entirely slanderous.

Here, then, was the background to Voss's trip to Belgium. It is true that other constructions can be put on it – that he lost money at the casino, for instance, or that his deal with M. Bogaert was a disappointing failure – but such constructions are not supported by the known facts. With his associate Dearden he then went to the airfield and took out a £500 insurance policy for the flight, a perfectly natural thing for a prudent man to do. Dearden did it as well. Surely if Voss had intended to destroy himself and the plane, some inkling of it would have dawned on Dearden.

To exonerate Voss from all blame, however, leaves one with an insoluble problem. If Voss was innocent, whose hand wrought the destruction of the *City of Liverpool*? Or was it an accident? It seems doubtful that so fierce and sudden a fire, in a comparatively invulnerable part of a well-proved airliner, could be accidental and leave no trace. There remains the question of sabotage.

The investigations showed that the fire started in either the luggage rack or the lavatory; suspicion centred on the luggage itself. The subsequent history of air sabotage shows that the luggage and the lavatory are the normal places of concealment for the mechanical time bomb.

The *City of Liverpool*, it will bc remembered, was 36 minutes late. Had she left Brussels on time she would have been over mid-Channel when the fire started. In this case, in all probability, nothing would have been known of her fate.

This is the dream of the air saboteur – to bring his prey down on the water, so that it disappears without trace. It becomes at once the perfect crime.

Who, in March 1933, determined to destroy the *City of*

Liverpool together with all her passengers and crew – and thus become the perpetrator of the first of that most hideous of all mass murders, air sabotage? What was the motive?

More than 30 years later, these questions still remain unanswered.

5

LESLIE HOWARD AND THE PARTY OF FOUR

THE main airport building at Portela, Lisbon, vibrated with the sound of aircraft engines. Outside on the tarmac apron a twin-engined DC3 transport, better known as a Dakota, with the code-letters G-AGBB, was running up its engines.

The passenger-handling staffs of the big airlines – BOAC, KLM, Lufthansa, Swissair – stared out of their windows for a moment and then bent over the desks of their offices fringing the high, echoing assembly hall. One of them, perhaps, reached for a telephone.

The date was 1st June 1943. Another wartime passenger flight was about to begin. This one was bound for the United Kingdom. Unarmed and unescorted, the plane was clearly marked with insignia and registration letters of the operating company. It belonged to the Royal Dutch Airline, or KLM. Several Dutch crews had escaped from Holland when the country was overrun in 1940, and three complete KLM crews were now operating a regular service between Lisbon and 'an airfield in Britain'. The security cliché hid the identity of a civil airfield at Whitchurch, near Bristol. But while precise details of the aircraft's route and destination could be kept secret, the point and time of departure at the Lisbon end could not. When the plane eventually took off, there was little doubt that further telephone calls would be made. Especially if it carried anyone designated VIP.

Portugal was a neutral haven in which diplomats of the warring nations rubbed shoulders and where spies and agents congregated, intermingled and plotted, and were periodically fished out of the Tagus. Mostly the diplomats ignored each other. But they could not altogether ignore the spies and agents. Every bartender might be listening for odd scraps of conversation as he mixed and shook his cocktails, every beautiful woman might be in the pay of at least one foreign power.

On board the Dakota were 13 passengers and a crew of four. Among the passengers were three women and two children. None of the passengers had any military status. The men were mostly diplomats or business executives. One was a correspondent for Reuters. Tyrell Shervington was one of the principals of Shell-Mex in Portugal; he was also suspected by German counter-agents of espionage activities. Berthold Israel was sponsored by the Colonial Office and had visited Lisbon on behalf of the Jewish Agency in Palestine. His mission was to help Jews who had escaped from Nazi persecution to reach Palestine. Gordon Maclean, Inspector of Consulates for the Foreign Office, had been visiting Lisbon and Madrid.

The last two passengers to board the plane were a Hollywood film star who had achieved world-wide fame, and his friend and business manager. The latter, Alfred Chenhalls, was a man of Churchillian visage and frame who sported a bowler hat and a large cigar. The film star was Leslie Howard. 'I'll do anything that I'm wanted to do,' he had said when, unlike so many of the British community in Hollywood, he had returned to Britain soon after the outbreak of war. And the British Council, with its mission of making the British way of life more widely known and understood abroad, had asked him to help.

For the past month he had been travelling in the Iberian peninsula, lecturing on aspects of the stage and cinema and showing propaganda films for the British Council. It had been an arduous tour, but it had been an enormous success. Over 900 cinemas in Spain and Portugal had agreed to show British documentary films. And one of the films he took with him, 'Pimpernel Smith', in which he played the part of a professor of archaeology who snatched victims of Nazi persecution from under the noses of the Germans, was voted by the Portuguese the finest film of the year.

How all this must have angered the Germans! Years of Nazi propaganda and vituperation were being eroded by this modest, self-effacing actor with the aesthetic features and the gently humorous smile. The British Council could have found no one who so personified the popular idea of an Englishman. Even the years in Hollywood had done nothing to debase that intensely English voice, appearance and manner. The vague, shy, self-conscious character that he assumed in many films was natural to him. Slight and a little stooped, with fair hair beginning to turn grey, he was happiest when, at his country home near Dorking in Surrey, he could relax and assume the garb of the week-ending Britisher – pipe, flannels and an old tweed coat.

Howard had planned to return to England on 29th May, but he was persuaded to remain in Lisbon for the première of his latest film 'First of the Few', in which he played R. J. Mitchell, the designer of the Spitfire. He made no secret of his intended departure, nor of the delay. Why should he? He was so modest about himself that it did not occur to him that anybody would be much interested in his movements. 'I hope to go back shortly to supervise the final stages of my film "The Lamp Still Burns",' he told a questioning reporter. His statement was widely reported in the world press.

The delay gave him a few days well-earned rest at Estoril, where he stayed at the Atlantico Hotel. 'Sitting by the side of the Atlantic Ocean,' he wrote to a friend, 'for the first time I am able to collect my thoughts.' He had become disillusioned with his life as a film actor and had talked of giving up acting and concentrating on directing, of which he had already done a good deal. He had refused to play in 'Intermezzo' (with Ingrid Bergman) unless he was made an associate producer. His one remaining ambition as a film actor was to play Hamlet, and one of his lectures in Madrid had been on 'The Modern Actor's Approach to Hamlet', a part he had played on Broadway.

He had no premonitions about the forthcoming flight. He had done a little shopping in Lisbon, buying presents for his family and for the girls on the set at Pinewood. Characteristically, because he was fond of children, his one regret of the tour was that he had been forced by his crowded itinerary to disappoint an audience of children at the British Institute in Madrid. Now, with a few hours to spare in Estoril, he wrote to the director of the Institute to apologize again and to ask if there was anything he could send the children from England to make up for it.

On the day of the 'First of the Few' première, the Director of Films for the British Council, a man named Neville Kearney, visited Howard and Chenhalls in their hotel at Estoril. Before lunch, in the cocktail bar, Kearney recognized an attractive Hungarian woman whom he suspected of being an enemy agent. He excused himself for a moment and checked at once by telephone with the first secretary at the British Embassy in Lisbon, who confirmed his suspicions.

Unknown to Howard and Chenhalls, they were being kept under the closest surveillance 24 hours a day. It had been the same in Madrid, where a woman whom Howard had befriended proved to be a leading enemy agent. The Germans just could not believe that an organization like the British Council could be anything other than a sophisticated Intelligence Service, with its members and lecturers nothing less than high-powered agents.

When Kearney returned to the bar he found Howard and Chenhalls just going in to lunch. Leslie Howard was talking about the return trip. 'We're going home,' he was saying, 'by the morning plane from Lisbon Airport on June the 1st.' Kearney was shocked to hear this said so openly, within earshot of the Hungarian woman, and later, when they were alone, he chided them both. But in any case it was clear that, the way the German agents were shadowing them, their departure would be no secret.

Early on the morning of 1st June, Howard and Chenhalls were driven in an Embassy car to Portela. In addition to the 13 people who comprised the final passenger list, one other person was preparing for the flight. His name was the Rev. Arthur S. Holmes, vice-president of the English College in Lisbon, who was returning to England on leave. Early that morning, Holmes called in at the College to collect some personal belongings. While he was there the telephone rang.

'Is that the Reverend Holmes?'

'Speaking.'

'This is the Airport. We're having a little difficulty over the seating on this morning's plane. Mr Leslie Howard is now travelling with a party of four, and at the moment we can find seats for only three of them. If it should be necessary I wonder whether, as a special favour to Mr Howard, you would mind if we put you on a later plane?'

'Not at all,' said the Reverend Holmes. 'Do you still want me to come to the airport?'

'Oh yes – we're still trying to fit everyone in.'

What was the meaning of this reference to a party of four? Were there two additional passengers, whom the airport staff were trying to squeeze in at the last minute? If so, who were they? A study of the passenger list suggests that the most likely pair to attach themselves to Howard and Chenhalls were the representatives of the Foreign Office and Colonial Office – Gordon Maclean and Berthold Israel, both of whom had met Howard and Chenhalls frequently at the Embassy in the previous few days. Another possibility was Shervington, the man whom the Germans suspected of being a British agent. But whoever they were, unless the airport staff had merely made up the story to satisfy the Reverend Holmes, it seems clear that Howard was travelling with a party of four. This may have no bearing at all on subsequent events, or it may be significant.

When Holmes got to Portela he was told that arrangements had been made to accommodate the additional passengers and that it would not after all be necessary for him to give up his seat. He boarded the plane with the other passengers. There was then a short delay while a box of nylon stockings which Howard was taking home as gifts, and which he had earlier deposited in bond, was recovered by an airport official. The official brought something else, in addition to the parcel. It was a message for the Reverend Holmes.

'Mr Holmes – you're wanted urgently at the English College.'

'Did they say what for?'

'No – just that you're wanted immediately.'

Holmes collected his luggage and left the aircraft. He would be travelling by a later plane after all.

The Dakota taxied out and lumbered off down the runway. Its route would take it along the Portuguese and Spanish coasts to Cape Finisterre and then out across the Bay of Biscay, keeping well clear of Brest, before turning east-north-east for Whitchurch. For more than half its journey it would be in danger of interception by German aircraft, mostly from long-range fighters protecting the U-Boat lanes in the Bay. But in three years of operating since Dunkirk, only two such attacks had been made. The first had been six months earlier, in November 1942, when the same Dakota that was now about to take off from Portela had been attacked by a single German plane about 250 miles from the English coast on its homeward run. The plane had been damaged considerably, but the pilot had managed to evade further attacks and gain the sanctuary of cloud. The second occasion had been only six weeks earlier, on 19th April 1943. This time the plane had been attacked from all angles by six Ju 88s in a most determined attempt to destroy it, and only superb airmanship by the Dutch pilot, Captain K. D. Parmentier, had averted disaster.

After this second attack it was seriously considered whether the scheduled flights between Lisbon and the UK would not be better undertaken at night. But there were operational difficulties, and no change had been made so far. Meanwhile, since news of the departure of the flights could easily be signalled by Lufthansa agents in Lisbon to Luftwaffe bases in France, daylight interception over the Bay of Biscay remained a serious hazard.

The Dakota G-AGBB took off from Portela at exactly nine-thirty, Double British Summer Time. Whether any long-distance messages signalled its departure to enemy airfields in Occupied France is not known. But one thing is certain. Half an hour later, at ten o'clock, eight Ju 88s of a fighter wing attached to the KG 40 bomber group took off from their airfield at Kerlin Bastard, near Bordeaux, and headed out across the Bay, on a course that would cross the estimated track of the Dakota. Their mission, ostensibly, was the protection of two U-boats that were traversing the Bay.

As the Dakota headed northwards along the Portuguese coast it ran into drizzling rain. It looked like being a typical Bay crossing. For much of the time the pilot, Captain Quirinas Tepas, was on instruments as the aircraft ploughed its way through opaque cloud. But on this route, no one minded bad weather. It greatly lessened the risk of interception.

The Dakota was a comfortable aircraft with a modern cockpit layout and the latest radio aids. It was well equipped for the worst weathers. And for the passengers, instead of the normal acute discomforts of wartime flying, there were luxurious tilted seats and a steward dispensing travelling rugs, coffee, drinks and food, and the latest German magazines, creating an illusion of peacetime air travel. Parachutes were not normally carried on these flights, and the only reminder of the dangers of a wartime ocean crossing was the life-jacket drill which preceded take-off.

The normal routine half-hourly exchange of signals, giving the aircraft's position by dead reckoning, was maintained by the radio operator, Cornelius Van Brugge. After two hours flying G-AGBB was off Corunna, about to begin the crossing of the Bay. From this point on, wireless silence would be observed, except in emergency. Nothing could be seen of the Spanish coastline and there was nothing unusual to report.

Meanwhile the eight Ju 88s from Bordeaux were continuing their patrol. Conditions were altogether unsuitable for search work due to low cloud and bad visibility, and they had failed to make contact with either of the two U-Boats. Yet they had not abandoned their patrol. Among their routine tasks, in addition to the escort of U-Boats, was air/sea rescue, but they had no such commitment today. Another was the interception of Allied anti-submarine aircraft operating in the Bay.

By 12.45 they had patrolled to a distance 500 miles west-north-west of Bordeaux, a few miles beyond the probable track of the Dakota. Now they were running back towards Bordeaux. Whether by accident or by design, they were swinging to and fro across the bows of the approaching Dakota.

And they had emerged into a patch of clear weather, where visibility had risen to four or five miles. It almost looked as though they had chosen to patrol in this clear patch to give themselves a better chance of making an interception.

The Dakota was at the limit of visibility when the first German pilot sighted it, broadside on. Soon afterwards Captain Tepas, peering through his wide windscreen into the mists of the Bay, saw a single aircraft far out on his port side. A moment later it turned towards him. He called Van Brugge.

'Get an emergency message out to base. Tell them we're being followed by an unidentified aircraft.'

Van Brugge, sitting at his radio immediately behind the second pilot's seat, made an immediate distress call on the control frequency, using only call-signs and a pre-arranged code-group. 'GKH from G-AGBB. An unidentified aircraft is following me.' He repeated the message several times, without receiving a reply. It was not heard at Whitchurch, but it was picked up in Lisbon. By the time it was received and understood, however, another code-group had been added by Van Brugge. The translation of this code-group was even more dramatic.

'I am being attacked by enemy aircraft.'

The eight Ju 88s had taken several minutes to form up and develop their attack. There was no question of a sudden trigger-happy burst of firing from a raw formation. These men were veterans, they had recognized the Dakota transport plane, and the attack was made in cold blood. All eight fighters swept down on the unarmed Dakota, firing as they came. Captain Tepas, trapped in the lane of clear visibility, had no hope of escape. He thrust the stick forward to try to make a run for it at sea level, but long before he could get there the Ju 88s had struck. As they broke away, the engines of the Dakota were streaking smoke and flame.

The fire in the engines quickly gained a hold, and as the incendiary bullets did their deadly work in the fuselage, the dive accelerated. The Dakota was out of control.

'Put on life-jackets. Fasten safety-belts.' Such orders surely must have been given. It is even possible that, for a special party of four whose safety may have been a matter of exceptional, national importance, and for whom special emergency precautions had therefore been taken, another order followed.

'Bale out. Bale out.'

For as the blazing Dakota plunged towards the sea, the German pilots, following it down, believed that they saw four parachutes flicker away from the fuselage door. The pilots were near enough to pronounce on the identity of the escapers. They were, they reported, four men.

Of the four parachutes, only two appeared to open properly. Of these two, one was already on fire.

The Dakota, now a mass of flames, dived headlong into the sea. As the German formation swept down after it, a pall of black smoke lay on the surface, and when they circled afterwards, nothing could be seen but floating wreckage. Even the single parachute which had apparently functioned properly had disappeared.

The eight Ju 88 pilots set course for Bordeaux, leaving any survivors who might be floating in the area to their fate. It seems odd behaviour for a squadron trained in air/sea rescue work. Yet it is fair to say that the pilots were convinced that no survivors existed.

When nothing further was heard of G-AGBB, a widespread search of the area from which its last message had emanated was begun. No trace of wreckage or survivors was found. Within 48 hours it was accepted that passengers and crew had perished. But for the brief message transmitted by Van Brugge, nothing would have been known of their fate. With this information, though, it seemed inescapable that an unarmed, unescorted transport plane, bearing clearly visible commercial markings and carrying civilian passengers only, including women and children, had been destroyed by enemy action. The suspicion was confirmed when the relevant German documents were captured after the war.

At the time of the incident, German propaganda described the aircraft as a British Douglas machine and claimed that such planes had been used as bombers – a claim that was without foundation. 'It had camouflage paint,' went on the apologia, 'and was in no way distinguishable from a war plane.' The incident had occurred over the Bay of Biscay, which was 'a zone of operations where British aircraft are waging war against U-Boats on their way to and from German bases'. Like most Nazi propaganda, it was a calculated mixture of lies and truth. Another propaganda ruse, employed without consideration of the effect it must have on the bereaved, was the release of a report that a Spanish fishing-vessel had picked up survivors, lessening the impact of the tragedy. The report was vigorously denied by the Spaniards, but it had its effect.

It is noteworthy that, in the records of the German squadron concerned, the Dakota is described as a courier or transport DC3. That the plane was plying between Lisbon and the United Kingdom on a legitimate commercial service must have been well known to the Germans. However, they hinted that the passengers were of military importance. Leslie Howard, they said, was a manufacturer of aircraft parts besides being a member of the British Intelligence Service. Shervington, of Shell-Mex, was the chief of the British Secret Service in Lisbon. And so on. Was there any truth in this?

It was natural that the Germans should look upon the British Council as a glorified intelligence service. The task it performed, although not linked to espionage, must often have fulfilled a similar function, promoting Britain's interests at the expense of the Axis. There remained the possibility that Leslie Howard was engaged on some special intelligence mission, for which his activities with the British Council were no more than a screen. Such a suggestion is utterly rejected by British official sources, but it would be a natural assumption from the German point of view. Credence was lent to the possibility by the cloak-and-dagger roles Howard had played in many films, and by the romantic nature of his life as a film star.

Why had it been so essential for the 'party of four' to travel together? And what, in the last plunge of the Dakota, was the identity of the 'four men'? Was there some reason why their escape was given priority over the women and children aboard? Or is it more likely – since none of the supposed parachutes seems to have functioned properly – that four passengers in lifejackets were hurled from the plane as it broke up and that the German pilots were mistaken? Passenger-handling staffs at Lisbon aver that parachutes were never carried on this service, and parachute-packs were bulky and not easily concealed. Outside the report by the German pilots, there is no evidence that parachutes were carried by Howard or his party or by anyone else on board.

There can be no doubt, in view of the earlier attacks on the same plane, and of the subsequent absence of any disciplinary action, that the eight men who shot down Dakota G-AGBB had orders to do so if they could. But did they have specific orders to seek it out and shoot it down on this particular day? The known movements and behaviour of the eight fighters, suspicious as they seem, could be coincidental. The evidence of the time of their take-off, and of their area of search, is circumstantial. There is no evidence of a direct order in captured German documents, no smug reference in a captured diary. If a secret order was given, that secret was well kept. And those Germans who might remember the incident are understandably reticent now. But let us suppose that the sequence of events began with a telephone call from the Lufthansa offices at Lisbon Airport, and that the Ju 88s were deliberately routed to intercept the Dakota. Who were the Germans after? Leslie Howard? Or were they, perhaps, after bigger fish still?

At the time of the tragedy, Churchill was attending a conference in Algiers with other war leaders to discuss the impending invasion of Sicily. This is what he said about the incident in his memoirs:

'As my presence in North Africa had been fully reported, the Germans were exceptionally vigilant, and this led to a tragedy which much distressed me. The regular commercial aircraft was about to start from the Lisbon airfield when a thickset man smoking a cigar walked up and was thought to be a passenger on it. The German agents therefore signalled that I was on board. Although these passenger planes had plied unmolested for many months between Portugal and England a German war plane was instantly ordered out, and the defenceless aircraft was ruthlessly shot down.'

Could Alfred Chenhalls have been mistaken for Churchill? Since Chenhalls himself had been under the strictest surveillance throughout the Howard tour, since he was a man 25 years younger than Churchill, and since Churchill was still at his conference in Algiers, it seems incredible that the Germans could have been so stupid. Churchill himself seems to have doubted it. 'It is difficult to understand,' he wrote, 'how anyone could imagine that with all the resources of Great Britain at my disposal I should have booked a passage in an unarmed and unescorted plane from Lisbon and flown home in broad daylight. We of course made a wide loop out by night from Gibraltar into the ocean, and arrived home without incident.' But this was five days later.

Churchill was in error in saying that the plane had plied unmolested; a most determined effort had been made to shoot it down only six weeks earlier. It may indeed be asked why, in the intervening weeks, the schedule wasn't altered to allow the crossing of the Bay to be made at night, whatever the operational objections. It had to be altered after the Howard incident.

If the Germans were out to get someone in particular, it must have been Leslie Howard. Neville Kearney, of the British Council, believed that Goebbels himself ordered the shooting, partly because of Leslie Howard's anti-Nazi films and partly for an imagined affront years earlier, when Goebbels quite wrongly believed that Howard had enticed away an attractive German woman in whom Goebbels was interested. And it might be naïve to dismiss out of hand the possibility that Howard had accomplished some secret Intelligence mission quite outside his work for the British Council. The invasion of Europe in 1943 was still a distinct possibility, and Spain was being wooed by both sides.

It has been fiercely denied that Leslie Howard was implicated in any Intelligence coup, as though that were something to be ashamed of. Surely the reverse is true. It is worth recalling that Military Intelligence has never been slow to exploit the talents of men of the arts; the undercover work of Compton Mackenzie and Somerset Maugham are pointers.

Mystery still surrounds the last-minute phone-call that saved the life of the Reverend Arthur Holmes. He went straight back to the English College, but no one there had any knowledge of the phone-call, or if they had they did not admit to it. The inference was clear: someone in Lisbon knew that Howard's plane was about to run into very great danger and was anxious that Holmes should avoid it.

Yet in spite of the many puzzling aspects of the tragedy, it may well be that the least spectacular explanation is the most likely one: that the regular schedules of the Lisbon-UK service had outrun whatever usefulness they had for the Germans; that they were regarded in Berlin as a valuable communications channel for the Allies; and that a decision had been taken weeks earlier – before the previous determined attempt to shoot the plane down – that the immunity which had earlier been granted to them must be ended.

If this is the case, the slaughter when it came was ruthless but impersonal.

6

THE LONG SILENCE OF *STAR TIGER*

THE fifteen men who gathered in the upstairs room in Grafton Street, Mayfair, in January 1946 were among the most highly decorated of the war. Nearly all of them had been members of that élite corps formed to find and mark targets in Germany for destruction by Bomber Command – the Pathfinders. Now they had exchanged their uniform, with its rows of medal ribbons and the coveted gilt Pathfinder wing, for a new uniform – that of the grey, chalk-striped demob-suit. Yet for these men the excitement and danger of the war years were about to be superseded by a fresh and stimulating challenge. Soon they would don the dark-blue uniform of an airline. They would pioneer new routes. One day, perhaps, they would become the custodians of the jet age.

For these men, as they pored over their charts in that upstairs room, plotting routes and practising astro-navigation prior to sitting for their flight navigator's licence, the problems of the return to Civvy Street were attenuated if not quite removed altogether. There was an adjustment to be made, but it was not fundamental. Flight planning, navigation, even the aircraft they would be flying, were not much different from the war years. They would be doing basically the same job. They would even have the same boss. The offices on the ground floor displayed the name of a new airline, whose charter had just been granted by Parliament. Its title was British South American Airways. The chief executive was Pathfinder Don Bennett.

Of the fifteen men who commuted daily in their demob-suits and RAF raincoats to and from that crowded upstairs room, thirteen are now numbered among the senior jet captains of BOAC. This is the story of the two odd men out. And especially of one of them.

Thirty-four-year-old Brian McMillan was a stocky, fair-haired, capable New Zealander who had come to England to join the RAF as a pupil pilot in 1937. He spent the first four years of the war in India and Burma, winning the Air Force Cross for his tireless work of evacuation during the retreat. Returning to England in 1943, he became one of that small group of airmen, the élite of the élite, entrusted with the role of master bomber, winning the DSO and the DFC. He had completed nearly 3,000 hours flying by the time he was demobilized. Steadiness and resolution rather than brilliance were his characteristics, and he was ideally suited temperamentally to airline work. It was natural, when the new Avro Tudor passed its trials and was introduced into service by British South American Airways, that McMillan should be one of the men specially chosen by Don Bennett to operate a new Transatlantic service from Lisbon to Cuba via the Azores and Bermuda.

Three of these Tudors, *Star Panther, Star Lion* and *Star Tiger*, were delivered by the manufacturers in the autumn of 1947, and for the next three months the service was operated safely – though not without incident. With the inauguration by BSAA of this new route, the leg from the Azores to Bermuda became the longest ocean crossing plied by commercial airlines anywhere in the world. Here, at the widest point of the great bowl of the Atlantic, lay more than 2,000 miles of featureless sea. Winds were westerly or south-westerly, mounting a frontal assault on planes westbound for Bermuda. And unlike the North Atlantic routes, there were no weather-ships dotted across this stretch of ocean; the requirement had been agreed but none had yet been provided. Commercial shipping in the area, too, was scanty, and no reliable data could be obtained from that source. Thus weather information was sketchy, and the status of the met. forecaster at the Azores little above that of tipster.

Theoretically the Tudor's endurance was adequate, but with adverse winds the margin could be narrow. Indeed, headwinds on this route were often strong enough to prevent a Tudor from reaching Bermuda at all. This presented no danger when the winds were correctly forecast; aircraft simply waited at the Azores for better weather. The real hazards came when winds increased beyond the forecast level during a flight. Twice in the first three months *Star Tiger* had been forced to divert to Newfoundland, and on another occasion a Tudor had crept into Bermuda with its fuel gauges showing empty after alerting the air/sea rescue services and preparing to ditch.

An acceptable safety factor therefore had to be injected into the technique of operating the route. Pilots planned their flights in three phases. Initially on the ground, and subsequently in the air, they computed a point of no return, the position of which depended on wind strengths, aircraft heights, and known fuel consumption and reserves. Beyond this point of no return they passed to the second phase, in which although committed to onward flight, they could still reach an alternative airport in Newfoundland if increased headwinds cut the safety margin for Bermuda below the agreed minimum. (Pilots were required to arrive at Bermuda with not less than two hours' fuel in hand, to cover the risk of sudden squalls which often blanketed the island for that length of time.) Finally, when Newfoundland passed out of reach, came the third phase, in which the plane was irrevocably committed to flying on to Bermuda, whatever the headwinds, whatever the weather. No other land could be reached.

All the hazards of this long and tiring flight between the Azores and Bermuda were well known to Captain McMillan when, on 27th January 1948, he took off from London Airport on another round trip to Cuba. The aircraft assigned to him for the first part of the trip was the Tudor *Star Tiger*. He would fly it as far as Bermuda, where a relief crew was scheduled to take over, leaving him to rest with his crew for a day or so before taking command of another westbound Tudor.

There were five other crew members in addition to Captain McMillan. The First Officer, Captain David Colby, a personable young man of great charm, was accompanying McMillan

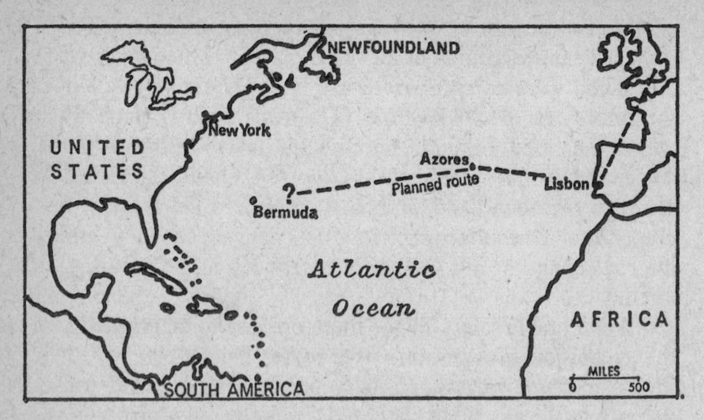

4. The Azores to Bermuda

on a familiarization tour before himself taking command of a Tudor. He already had over a thousand hours in command of BSAA aircraft, and like McMillan he was an ex-Pathfinder and had been one of the original fifteen in the upstairs room in Grafton Street. The best-known character amongst the crew, however, was the radio officer, Robert Tuck, better known as 'Tucky', whose effervescent spirits, comical asides and malleable, india-rubber face were familiar to all Trans-altantic fliers. Tucky, the oldest man in the crew, had had 14 years as a sea-going operator before transferring to airways work, and he had been Senior Radio Officer at the Prestwick end of the Atlantic Ferry throughout the war. He brought hilarity into any group he joined, and his job was the one thing he took seriously. Completing the crew were the Second Officer, Cyril Ellison, another ex-Pathfinder man, and two air hostesses, known as 'Star Girls': the dark-haired, black-eyed Lynn Clayton, survivor of a fatal York crash at Dakar nine months earlier; and Sheila Nicholls, a blonde whose wonderfully fresh complexion made her look younger than her 24 years. These two girls had 25 passengers to look after, among whom the most distinguished was Air Marshal Sir Arthur Coningham, formerly leader of the Desert Air Force and commander of the Second Tactical Air Force during the invasion of Europe.

The first leg of the trip, the flight to Lisbon, was uneventful except that the passengers suffered acute discomfort through a failure in the heating circuit – not an unusual occurrence in the Tudor at this time. Icicles formed in the roof of the cabin due to condensation. 'I am trying to write this letter,' wrote one of the passengers to a friend, 'at 21,000 feet over the Bay of Biscay. We cannot see anything because we are in thick cloud and the windows are frosted over inside. The heating has broken down and the thermometer is reading 34°.' The crew were also having trouble with one of the compasses. These defects were remedied during the night-stop at Lisbon, but trouble with the port inner engine delayed the take-off next morning. *Star Tiger* eventually got away from Lisbon at 11.45, two and a half hours late. Almost at once the heating circuit failed again. So, too, did the troublesome compass.

Two other BSAA pilots had a business interest in the progress of *Star Tiger*. The first, Captain Frank Griffin, a cheerful, phlegmatic Devonian, was waiting at the Azores for favourable weather to fly a Lancastrian with a load of freight to Bermuda, where it was to be transhipped to *Star Tiger*. The second, Captain Geoffrey Rees, was pilot of the relief crew waiting at Bermuda to fly the plane on to Cuba. (Griffin and Rees, too, were ex-Pathfinders who had been among the original fifteen in the upstairs room in Grafton Street.)

It was mid-afternoon when McMillan and Colby, in the cockpit of the Tudor, picked out the steep, rocky outline of Santa Maria, the most southerly island of the Azores group. It was here, during the war, that the British had built the airfield. Captain McMillan landed the Tudor into a 60-knot wind which brought the plane to a halt in less than 200 yards. According to his schedule he should be taking off again for Bermuda immediately after refuelling. But unless this wind was local there would be little chance of pressing on to Bermuda that night.

Flights between the Azores and Bermuda were always undertaken at night, because of the crews' dependence on astro-navigation on this leg. Take-offs were usually timed for the middle or late afternoon, so that darkness would overtake the plane two or three hours after take-off. The essence of the first phase was to have enough darkness to take a series of astral fixes before the point of no return, to establish the wind strength and relate it to the fuel position, so that a decision to continue or discontinue the flight could be made in good time.

After taxying to the control building, the crew of the Tudor walked across the tarmac to the BSAA flight office, where Captain Griffin and his crew were waiting for them. British South American Airways was a comparatively small concern, in which everyone knew everyone else. After an exchange of greetings, Griffin took McMillan and Colby along to the meteorological office. 'There's a strong wind,' he told them. 'I expect I can make it in the Lancastrian, but I've got long-range tanks and an endurance of 19 hours. I don't know about you.' The Tudor was slightly faster than the Lancastrian, but it consumed more fuel.

The three men studied the charts together. The powerful winds were increasing with height, and the only possible solution was to fly at low level. But thick layers of strato-cumulus were expected to obscure the sky for a thousand miles or more, making a position check by astro before the point of no return impossible. Griffin could probably have made a safe attempt in his Lancastrian, but there was no point in his getting to Bermuda a day ahead of the Tudor. The two captains decided on a 24-hour postponement.

When their decision was known, night-stopping baggage was off-loaded from the Tudor and the passengers were checked in and accommodated in the airport hotel. This was a single-storey, prefabricated building which consisted of a lounge, bar and dining-room, which led off to rows of small bedrooms on either side of lengthy corridors. After claiming their luggage the passengers congregated in the lounge for tea.

Both crews were moderate or light drinkers, but most of them sat up at the bar for a drink before dinner. The talk was mainly of domestic matters – rationing in England, petrol coupons, the housing shortage. And the war. Always in those days the talk came back to the war. Soon after dinner, however, Captain McMillan excused himself. 'I'm very tired,' he told Griffin. 'If you don't mind, I'm off to bed.' Griffin, who had done no flying that day, didn't feel like sleep just yet, and soon he was chatting with one of McMillan's passengers, an ex-RAF man named Tony Mulligan whom he had known in his Pathfinder days. After the usual recital of what had happened since the war to various friends and acquaintances, Griffin was introduced to Mulligan's wife and father, who were also travelling on the Tudor. All three were bound for Bermuda. Suddenly a thought occurred to Mulligan.

'I say,' he said, 'wouldn't it be marvellous if I could come with you tomorrow? It'd be great to fly in a Lanc again, even thought it's a Lancastrian now. I'd like my wife to see it, too. Could you take us?'

Griffin hesitated. He would have liked to take Mulligan and his wife along, but his Lancastrian was a freighter. It was not insured to carry passengers. It would be all right unless something went wrong. Then there would be a mighty row. 'I'm sorry,' he said. 'It's more than I dare do.'

Mulligan looked crestfallen, and later he returned to the subject again, but Griffin still reluctantly declined. Mulligan, his wife and father remained on the Tudor. It may be that Mulligan had some sort of premonition. If so, he kept it to himself.

A Mexican passenger, one of two who had boarded the plane at Lisbon, was writing to his wife. He had been told of a possible diversion to Newfoundland and the United States. 'The weather is still bad,' he wrote. 'We do not know if we will leave tomorrow or next day by way of New York. When we came here we were being blown by a 60-mile-an-hour wind and we were about to be blown to Canada or down into the ocean.' This perhaps was a pardonable exaggeration; the passengers had had an uncomfortable ride so far.

Early next morning McMillan, Griffin and Colby went to the forecaster's office to study the latest reports. The weather was much more favourable. The winds as always were westerly for much of the way, and they increased with height, but an acceptable headwind, which would give both aircraft the required safety margin at Bermuda, could be found at 2,000 feet. There would be a heavy overcast along the route, especially for the first 800 miles, but it was expected that star shots would be obtainable well before half-way.

'It looks all right to me,' said McMillan.

'Me too.'

'Right. We'll go this afternoon.'

The two crews carried out their flight planning together. It was decided that Griffin should leave that afternoon at half-past two Greenwich Mean Time (half-past twelve at the Azores), about an hour before McMillan, giving a time separation for handling at either end. Also Griffin, with the greater range, would be able to pass a warning back to McMillan by radio of any marked deterioration in the weather. A flight forecast for the route was prepared, and Griffin collected his copy at two o'clock before going out to the Lancastrian. As he left the control building he had a final word with McMillan. 'We'll keep in touch on the way,' he said. 'See you in Bermuda.'

For Captain McMillan there was a problem of loading. With 25 passengers and full petrol tanks his Tudor would be very slightly overweight. He was carrying no freight, so he was left with the choice of leaving two or three passengers behind or of taking on a slightly reduced fuel load.

'What will you do, Captain McMillan?' asked the traffic assistant.

'I'll reduce my petrol load by 150 gallons.'

But when it came to giving instructions for the reduced fuel load, Captain McMillan hesitated. From the economics standpoint he had been reluctant to disembark passengers. From the safety standpoint, the important thing was fuel. All past experience on this leg of the route emphasized the importance of every single gallon. In any case he knew that the Tudor IV, the type he was flying, was quite safe with a 1,000-lb overload. Most of this overload would disappear anyway with the consumption of fuel during taxying, and in the take-off run.

'Fill her up to the gills,' he told the ground engineer.

Star Tiger took off without incident soon after half-past

three, climbed to 2,000 feet, and set course. Her endurance was 16 hours. Captain McMillan expected the flight to take just under 12½. That left an ample margin for unexpected head-winds or a sudden deterioration at Bermuda.

The crew of *Star Tiger* were soon hard at work. There was no relaxing on this long and critical flight. McMillan and Colby were busy with the flying of the aeroplane, the checking of gauges, and observation of the weather. Ellison was using the last hours of daylight to take drifts. 'Tucky' was checking his radio equipment and establishing contact with Santa Maria. It would be a long time before he could work Bermuda. At the rear of the cabin, Star Girls Lynn Clayton and Sheila Nicholls were preparing the evening meal.

The weather was as forecast, an active cold front bringing an impenetrable bank of strato-cumulus at 2,000 feet, reaching down in patches of squall to the grey, wind-lashed water. Turbulence was severe, and some of the passengers felt queasy, but they were more comfortable than they had been so far because heating was unnecessary at this height. At four o'clock Tuck transmitted the first of the hourly position reports given him by Ellison, computed entirely from dead reckoning. It might be many hours before they could take a star shot.

McMillan called Tuck on the inter-com. 'Give Frank Griffin a call. Ask him what wind he's getting.'

At eleven minutes past four the two aircraft made their first radio contact. Griffin was experiencing a wind considerably in excess of that forecast, 55 knots against 30–40, and this information was passed and was duly noted by Captain McMillan. Soon afterwards, Griffin addressed a message to New York and Bermuda, via the Azores, asking for a wind forecast for the later stages of the flight and a landing forecast at Bermuda. Nothing was forthcoming in the next hour, and with the sky still overcast, Griffin became uneasy. Then, just before six o'clock, his radio operator made contact with the American base at Kindley Field, Bermuda, and passed the request direct to them.

'Landing forecast and upper winds urgently required.'

This information was just as urgent for the Tudor, which was flying some 200 miles behind the Lancastrian but overtaking it slowly. Both aircraft were working on the same frequencies and Tuck was able to receive all messages passed to and from the Lancastrian.

The answer from Kindley Field estimated the winds along the route as stronger than those forecast by Santa Maria. They were, however, dying out in strength on the approaches to Bermuda. On the basis of these winds and his experience so far, Captain Griffin put back his estimated time of arrival at Bermuda by one hour. He was still within his own safety limits, but he decided to confer by radio with Captain McMillan.

'My flight time on revised winds is 13 hours 28 minutes. What do you think?'

McMillan agreed and amended his own flight plan accordingly. A delay of one hour for the Tudor, although more serious, was still within acceptable limits. But the margin was beginning to narrow.

After five hours' flying 'Tucky' established contact with Bermuda. But it was not until *Star Tiger* had been airborne for over six hours and was almost half-way that a clear patch in the overcast enabled Ellison to take a star shot. His calculations confirmed the general suspicions about increasing headwinds. However, if the forecasts Tuck was now receiving from Bermuda were approximately accurate, the duration of the flight would still be under 14 hours, leaving the prescribed two-hour safety margin. Thus, an hour or so later, *Star Tiger* passed the point of no return and became committed to onward flight.

Soon after midnight the Tudor reached a position on a line 50 degrees West, 750 miles from Bermuda. For two hours the overcast had closed in and astro-navigation had been impossible. The point of no alternative – where the final decision had to be made whether to press on for Bermuda or divert to Newfoundland – was fast approaching. So too was a dramatic change in the wind.

Shortly before one o'clock, when Ellison got his first astro-shot for three hours, there was no indication of it. But half an hour later, when Griffin's navigator, 150 miles further on, got his first astral fix for some time, he found that the Lancastrian was a long way off its course. According to his calculations the plane had been blown 68 miles north of its intended track in the previous hour. This indicated an unforecast south-westerly wind of about 50 knots instead of the light and variable winds previously forecast for this zone.

In this period, while the crew of the Lancastrian were awakening to the sudden wind-change and the crew of the trailing Tudor were still unaware of it, *Star Tiger* passed her point of no alternative, the point of final decision. The haven of Newfoundland thus slipped out of reach. Captain McMillan was committed to maintaining his course for Bermuda.

Griffin radioed news of the wind-change to McMillan, and twenty minutes later, at two o'clock in the morning, Cyril Ellison managed to get another astral fix. It revealed that *Star Tiger*, too, had been blown well north of her intended track. For nearly an hour, the Tudor had been crabbing away from Bermuda in the direction of New York. Ellison at once gave McMillan a new course to steer for Bermuda. It turned the Tudor right into the teeth of the gale.

Star Tiger had now been airborne for 10½ hours. She had about 550 miles to go. With the existing headwind the plane would not reach Bermuda for 4½ hours. But that still gave McMillan a precious hour in which to cope with the unexpected. If there was no sudden deterioration at Bermuda, he would still have a reasonable safety margin.

At three o'clock, with cloud once more obscuring the stars, *Star Tiger* reported her position by dead reckoning. The crew knew they were still well north of their intended track, but they believed they were making good their course for the island. Ahead of them, Captain Griffin in the Lancastrian was about an hour out of Bermuda. He radioed a message back to McMillan.

'Changing now to voice telephony to contact Bermuda Approach Control. See you at breakfast.'

The situation in *Star Tiger*, although not critical, was

precarious. The normal safety margin for a landing at Bermuda had been halved, and might be further reduced as the flight neared its climax. It was over an hour since the crew had been able to fix their position by the stars. Unknown to them, thick cloud covered the approaches to Bermuda and the chance of further star-shots was remote; in these conditions, beset as they were by capricious and violent winds, accurate navigation depended on radio.

'Get me a bearing from Bermuda, Tucky.'

So long as they were in radio contact with Bermuda, the situation was under control. Tuck requested a bearing immediately after sending the three o'clock position report. But *Star Tiger* was still nearly 400 miles from Bermuda, and the ground operator was unable to take a satisfactory bearing on her signals. He asked Tuck to call him again in a few minutes.

The pressure on Tuck to get a good bearing was mounting. At a quarter past three he called Bermuda again. This time his signals were received at sufficient amplitude to enable an accurate bearing to be taken.

'Your bearing from us is 72 degrees. Class 1.'

A Class 1 bearing would be no more than two degrees out. Any doubts Captain McMillan had about his position would now be quickly dispelled. A series of these bearings would guide him safely into Bermuda. But for the moment it seemed that he was satisfied. He was on course, he had his first-class bearing, and he had his safety margin, reduced but adequate. The ground operator at Bermuda heard nothing from him for the next half-hour.

There was nothing unusual in this. But strictly speaking, the ground station were supposed to make contact with aircraft under their control at least once in each half-hour. After 35 minutes the operator called *Star Tiger*. There was no reply.

Probably the Tudor had changed frequency to voice telephony for the last stage of the flight, as the Lancastrian had done. The operator rang Approach Control and asked if they had made contact.

'Nothing from him so far.'

Seven hours had elapsed since the ground operator had first

accepted control of the Tudor. In that time, silences of up to an hour had not been uncommon. He called the plane again soon after four o'clock, but there was still no reply. He had other traffic to deal with, and that absorbed him for a time; he did not regard the Tudor's silence as significant. It was not until twenty to five, 95 minutes after his last contact with *Star Tiger*, that he raised the alarm. By that time Captain Griffin had landed safely and was having breakfast with his crew in the American PX. It was there that he learned that the Tudor had failed to answer signals and was now overdue.

Working back to Tuck's last message, it was clear that something had interrupted radio contact shortly after the acknowledgement of the bearing. Perhaps there had been a radio failure. That seemed the most likely explanation. Other possibilities of a more catastrophic kind began to be canvassed as the Tudor's known endurance was dissipated with the further passage of time.

Geoffrey Rees, captain of the relief crew waiting at Bermuda, assumed command of the Lancastrian brought in by Griffin and took off on a parallel search soon after dawn. Griffin himself, although very tired after eighteen hours on duty, fourteen of them in the air, insisted on going with him, to provide another pair of searching eyes. The American search and rescue section at Kindley Field was also alerted, and a Fortress equipped with a radar scanner took up the quest.

In spite of appalling weather which at times rendered searching operations hazardous in the extreme, 25 other planes took part in the first day's search, and nearly a thousand hours of searching were crammed into the next five days. Not a trace of wreckage, not a patch of oil, not a dinghy, not a single body, was ever found. Just at the moment when all her difficulties seemed to have been resolved, *Star Tiger*, together with her complement of passengers and crew, had disappeared without trace.

Captain Griffin would testify that in his own experience, an hour ahead of the missing plane, there had been no turbulence at this stage of the flight, no icing, no fog, no electrical storms, nothing that could account for the aircraft's destruction.

Only a few minutes before the plane's last signal he had been in contact with it. Had there been any doubt in Captain McMillan's mind at that stage of his ability to reach Bermuda he would certainly have communicated it to Captain Griffin.

The most searching analysis failed to disclose any weakness in the construction of the Tudor which might account for the disaster, and the impression left in the public mind was that man and machine had been overwhelmed by some unknown phenomenon, natural or mechanical. This impression was not altogether dispersed by the subsequent enquiry, which eliminated radio failure, constructional defects, meteorological hazards and engine failure as likely causes and seemed to point a doubtful finger at either fire or mechanical disruption.

There was a mass of evidence about a faulty setting of the fuel cocks causing fuel starvation, and on possible reasons for loss of pilot control, but none of these could have caused the disaster unless McMillan and Colby had been negligent beyond the realms of credibility.

Why, however sudden the catastrophe, had the alert, experienced Robert Tuck been unable to send a distress signal? There were many stations listening on his frequency; had he as much as touched the transmitting key, there can be no doubt that he would have been heard.

Although not carrying radar, *Star Tiger* was fully equipped with the best radio aids of the day; the possibility of a complete radio failure was remote. Yet so were all other possible solutions. Leaving aside fanciful theories, the most likely failure would surely seem to be electrical.

As the reconstruction of *Star Tiger*'s flight shows, Captain McMillan had reached the stage when a safe arrival depended entirely on the correct functioning of his radio. Fuel was short and his position uncertain. Bermuda was a morsel of land in a desert of sea; he would have to be led to it by radio. That he recognized this is evident from the requests for a bearing, beginning some two and a half hours before he could expect to reach the island. If his radio failed his chances of finding Bermuda would deteriorate to thousands to one.

Where, then, did *Star Tiger* spend her last desperate minutes – or hours?

Sifting through a heap of unconfirmed reports of the sighting of dinghies and wreckage, and even of the plane itself, one eye-witness account stands out for its vividness. Somehow it lingers in the mind. It came from the look-out of a Norwegian ship, who believed he saw a low-flying plane about an hour before dawn, shrouded in darkness, 300 miles north-west of Bermuda. If this was *Star Tiger*, she was well off her course, and far beyond the haven for which she was making. Yet this was the direction in which the winds had been driving her.

The look-out's brief glimpse could have been man's last sight of the ill-fated *Star Tiger*.

7

RAY HINCHLIFFE AND ELSIE MACKAY

DINNER at the eighteenth-century George Hotel at Grantham on Monday 12th March 1928 was typical of the sort of meal served at that time at any provincial three-star hotel. The soup was thick and hot and appetizing. The fish had been landed at Grimsby that morning. There was the choice of roast beef, roast lamb, poultry or game. Glistening cutlery reposed with mathematical precision on the starched linen tablecloths. Tail-coated waiters glided silently across the thick pile of the carpet. The table in the corner was neatly laid for three.

By half-past eight that evening the dining-room was full. At the corner table a well-dressed woman sat between two men. All three looked to be in their early or middle thirties. Two of them – the woman and one of the men – were about to embark on yet another Atlantic air attempt from east to west. All previous attempts had ended in disaster. Three such attempts had been made in the previous year, costing seven lives, including those of two women. If this attempt was successful it would be the first time that a woman had flown the Atlantic.

The conversation at the corner table was animated but subdued. There was an atmosphere of conspiracy. The woman at the table was the Hon. Elsie Mackay, daughter of Lord Inchcape, then chairman of P and O. Virtually a millionairess in her own right, she was potentially one of the richest women in Britain. Now 34, she had been one of the first to take full advantage of the emancipation of women. Dark-eyed, black-haired and petite, she had defied her parents by choosing a stage career, and at 20 she was the youngest leading lady in London. Secretly married in 1917 to an actor whom she divorced five years later, she took up flying between stage and screen roles and qualified for her licence in 1922, one of the first women to do so.

It was in 1927 that she conceived the ambition of being the first woman to fly the Atlantic, and from the outset she realized that absolute secrecy was essential. Otherwise she knew that her family, and particularly her father, would do everything they could to restrain her. After consulting friends at the Air Ministry who were sworn to secrecy, she was put in touch with an Imperial Airways pilot named Captain W. G. R. Hinchliffe. Ray Hinchliffe at this time was probably the most experienced pilot in Europe, possibly in the world. 'Even among the best airmen of the day,' wrote Harry Harper, Britain's first air correspondent, 'he stood supreme.' Hinchliffe had learned to fly in the RFC in 1916, and he was such an exceptional pupil that he was employed at once as an instructor. Later he went to France, winning the DFC and the AFC. 'His handling of a Sopwith Camel,' wrote a contemporary, 'was the best I was ever privileged to see.' No one could imagine that he would ever crash or be shot down, but in May 1918, when flight commander of No. 210 Squadron, he was asked by his squadron commander to attempt the recovery before nightfall of a Camel which had been force-landed by another pilot in a field of growing corn. Hinchliffe made the attempt, but the plane somersaulted and he was badly hurt, losing the sight of one eye. The injury so disfigured him that he always wore a patch over the damaged eye. Yet his flying ability was not affected.

After the war Hinchliffe became the first chief pilot of the newly formed Royal Dutch Air Line (KLM), and he later transferred to Instone Air Lines and subsequently to Imperial Airways. He was one of a dozen pioneers of British civil aviation which included men like O. P. Jones, C. D. Barnard, Freddie Minchin, Gordon Olley, Donald Drew and Alan Cobham. By 1927 he had completed nearly 9,000 hours in the air. But now, attracted by the glamour of record-breaking, he began to tire of the monotony of civil flying. When the American Charles Levine announced his intention of attempting an Atlantic crossing from east to west and advertised for a pilot, Hinchliffe applied and was accepted. As a special concession he was given leave by Imperial Airways.

The project eventually fell through, but not before a proposal had been made by Levine that an American heiress named Mabel Boll should accompany them and thus become the first woman to fly the Atlantic. Hinchliffe was dead against it. 'If she goes,' he said, 'I don't.' There was nothing personal about it. 'A woman,' he said, 'is utterly unsuited to such a flight.' How was it that, six months later, he had completely changed his mind?

The answer lay first in the personality of Elsie Mackay. And the proposition she put to him was an attractive one. In the congenial atmosphere of lunch at the Ritz she offered him a salary of £80 a month plus generous expenses if he would buy a plane on her behalf and attempt the east-west crossing with her as second pilot. Fame would be his if he succeeded, not merely for the first east-west crossing by air, but for chaperoning the first woman across the Atlantic. It would be a double British triumph. She would foot all the bills, and any monies that might accrue from the flight – and several substantial money prizes were on offer for Atlantic and long-distance flights – would be his.

Hinchliffe had married a Dutch girl in 1922 and they had one child. A second child was expected before Christmas. Although still only 33, he knew that he could not go on flying indefinitely; for some time he had been worried about his sight. After long hours of route flying he invariably suffered from eye-strain, and he knew that sooner of later he would have to give up. What would happen to him then? Here was a chance, while he was still at the height of his powers as a pilot, to bring off a single ambitious stroke which might ensure his future. And if anything went wrong? Well, his family would be provided for. It was understood that Elsie Mackay, as part of the bargain, would insure his life for £10,000.

Hinchliffe made his plans for the flight with the utmost care. Finding that no suitable plane was available in Britain, he got leave again from Imperial Airways and went to America, where he bought a standard Stinson Detroiter monoplane and shipped it to England in the *Aquitania*. This type of machine was already in commercial use in America. The cockpit gave dual control and the cabin was roomy and seated seven passengers. The engine was the well-tried Wright Whirlwind, developing 200 hp and giving a top speed of 130 mph.

The machine was assembled at Brooklands aerodrome and fitted out for long-distance flight. After careful calculation Hinchliffe decided that with a crew of two the Detroiter could lift a total of 480 gallons of fuel, giving a duration of 40 hours and a range of 4,600 miles. Its normal capacity was 180 gallons. The cabin seats were dispensed with and a large tank holding 225 gallons was installed immediately behind the cockpit. The remaining 75 gallons were carried in 17 specially-made aluminium cans each holding 4½ gallons.

The stacking of these cans was a difficult process. Room had to be left for a passage-way through the cabin, not only for the crew to enter but to allow them to pass back through the fuselage and fill the cabin tank from the cans in flight. Another problem was that the carburettor was gravity-fed from the wing tanks, so that all petrol in the cabin tank had first to be pumped into the wing tanks. Two hand pumps and a mechanical pump were fitted for this purpose, and the whole intricate fuel system provided a maze of arterial piping in the cabin.

Yet another difficulty was to find a runway long enough for a fully loaded take-off. The best and longest runway known to Hinchliffe was at the RAF College at Cranwell. But after the tragedies of the previous year, public and official opinion viewed Atlantic flights with disfavour. The east-west crossing, involving as it did a battle against the prevailing winds, was especially unpopular. Despairing of getting official permission to make Cranwell his starting-point, Hinchliffe put the problem to Elsie Mackay. She went secretly to Sir Samuel Hoare, then Air Minister, and got his agreement to their use of the field at Cranwell. It was stipulated, though, that they must not be there for more than a week.

Meanwhile Ray Hinchliffe's ambitions had inevitably

clashed with his job. Imperial Airways had been sympathetic, but they could not grant him indefinite leave. In January 1928 he resigned from Imperial Airways and accepted the post of personal pilot to Elsie Mackay.

At last, on 24th February, the modified Stinson Detroiter, now christened *Endeavour*, was ready to fly to Cranwell. Painted jet black, but with its wires and struts picked out in gold, it was an impressive sight as it landed at Cranwell and taxied across the smooth grass field into the hangar. In the right-hand seat, where he had always preferred to fly since the loss of his left eye, was Ray Hinchliffe. In the left-hand seat was Elsie Mackay.

With the arrival of the plane at Cranwell the newspapers began to ask questions. The use of the long runway at Cranwell presupposed a heavily loaded take-off. Was Hinchliffe going to have another crack at the Atlantic? And, bigger news still, was Lord Inchcape's daughter going with him? Hinchliffe had announced early in February that he intended to attempt a new long-distance record, for which a prize of £5,000 had been offered; at Cranwell he told the Press that he was preparing for a non-stop flight to India. He denied any intention of flying the Atlantic. Miss Mackay, he explained, had a 'small financial interest' in the flight, but she would not take part in it. His companion would be an old RAF comrade, Captain Gordon Sinclair.

'Where is he?' asked the suspicious Press.

'He is indisposed,' answered Hinchliffe. 'Otherwise he would be here. Meanwhile Miss Mackay is accompanying me on my tests.'

Hinchliffe's known aversion to taking a woman with him six months earlier lent some credence to his cover story now. But the newspapers remained unconvinced. Then Captain Sinclair arrived at Cranwell, and Hinchliffe announced their imminent departure for India. Unfortunately at this stage the Atlantic weather deteriorated, making departure on the actual flight impossible. Every day of delay increased the danger that the cover story might be broken.

Then came another misfortune. Through a misunderstanding, Hinchliffe's agent in New York released the news that Hinchliffe was about to embark on an Atlantic flight, and that he planned to make for Philadelphia non-stop in an attempt to win the Atlantic award and the long-distance prize in one go. American newspapers splashed the story, and in Britain the *Daily Express* seized on it and added the important rider that the Hon. Elsie Mackay would accompany Hinchliffe and thus become the first woman to fly the Atlantic.

The report brought indignant denials. 'I am very annoyed about the whole matter,' said Elsie Mackay. 'I have no intention of accompanying Captain Hinchliffe.' Hinchliffe too issued a vehement rebuttal. 'The suggestion is ridiculous. Captain Sinclair is here and will accompany me. Miss Mackay will take no part in the flight.'

These denials silenced the newspapers for the moment. But the reports reached Lord Inchcape in Egypt, where Lady Inchcape was recuperating after a heart attack. Lord Inchcape at once wired his son: if there was any truth in reports that Elsie intended to attempt an Atlantic flight, she was to be dissuaded from doing so.

It was a nerve-racking weekend for Ray Hinchliffe and Elsie Mackay. The weather showed no signs of improving. Heavy falls of snow carpeted the airfield and the surrounding countryside. A telegram from the Air Ministry, reminding them that the agreed seven-day period had expired nearly a week ago, required them to remove their aeroplane from Cranwell forthwith. News came through that preparations for a German attempt on the same flight were well advanced. And during the weekend a deputation from the Inchcape family, consisting of Elsie Mackay's brother and brother-in-law, arrived at the George Hotel at Grantham in determined mood.

What passed between brother and sister was never revealed. No doubt Lord Inchcape's son stressed that the shock to the family if anything went wrong might prove fatal to Lady Inchcape. It seems unlikely that Elsie Mackay would have lied to her brother. Almost certainly she admitted what must have been obvious to anyone as close to her as her brother – that she had helped to plan the flight, had financed it, and fully intended going on it. It also seems probable that her brother, while perhaps unable to persuade her to give up the idea altogether, extracted a promise from her that in view of her mother's illness she would reconsider her decision to go.

Hinchliffe must have known all this. He was now in the position of a man who had burned his bridges, staking all on a single throw. He had given up his job. Much as he might sympathize with Elsie Mackay if she showed signs of weakening, any further delay would be disastrous for him. Already there was the threat that the Germans might beat him to it. Whatever happened he must go ahead with his own plans as quickly as possible. A man as powerful as Lord Inchcape might yet find some way of stopping the flight.

The weather charts on Monday, 12th March, showed only a very slight improvement. But the authorities at Cranwell insisted that he must leave the airfield at the very latest next day. Hinchliffe, like many fliers of his time, was extremely superstitious. He would not sleep in a hotel bedroom numbered 13. Tomorrow was Tuesday the 13th. Yet there was no other airfield from which he could fly off his fully-loaded machine. There remained the imminent threat from the Germans and from Lord Inchcape. He decided that the time had come.

It is probable that no other flier would have considered making such a flight in the prevailing weather conditions. But Hinchliffe was unique. For years he had operated daily air services through the worst weather in Europe, and the Atlantic held no terrors for him. He had never yet encountered weather that he could not fly through.

In all other respects, too, Hinchliffe regarded the flight with perfect equanimity. He had been preparing for it for nearly twelve months. He had studied the prevailing weather conditions and winds in the Atlantic at this time of the year and carefully plotted his course. He had personally inspected and chosen the plane, and supervised its assembly. In a long series of proving flights he had ascertained the plane's capabilities and computed its fuel consumption and range. He had replaced the aircraft compass with a special new type, and he had fitted a drift indicator. 'My confidence in the success of the venture,' he wrote that evening in the aircraft log, 'is now 100 per cent.'

And so to dinner in the George on the night of the 12th, with Gordon Sinclair now substituted for Elsie Mackay as co-pilot, after a tremendous struggle in that young woman's conscience. After dinner Hinchliffe drove back to Cranwell with his mechanic to arrange for his plane to be ready for an early take-off. 'I am definitely going tomorrow,' he told a reporter, 'and Captain Sinclair is coming with me.'

'What flight do you mean to attempt?'

'Really I don't quite know yet,' said Hinchliffe, feeling that an evasive answer might still serve him best. 'It all depends on the weather report I shall get in the morning.'

It was after midnight when Hinchliffe returned to the George. Elsie Mackay was supervising the preparation of sandwiches and coffee. Hinchliffe asked for a call at 5 AM. There was little sleep for either of them – especially for Elsie Mackay. All night she must have been plagued with doubts. Had she been right to allow herself to be talked out of her one remaining ambition? Was it selfish of her to want to go, or selfish of her family to want to stop her? The more she thought about it the more she must have been inclined to think the latter. Would it not be morally right, both for her and ultimately for her family, that she should go through with what she had set out to do, whatever the cost? Wasn't that the hub of the whole matter?

She had complete confidence in Captain Hinchliffe, and in herself. In 36 hours the flight would be successfully completed and she would never forgive herself for backing out. There was only one doubt left in her mind. After all that had happened, would Captain Hinchliffe take her?

At 5 AM she was awakened by her maid. She put on her leather flying tunic and trousers and then covered them fairly effectively with her fur coat. She pulled on her flying helmet and wound a silk scarf round her head to hide it. Then she watched at the window of her room, waiting for her limousine to pull in outside the hotel. As soon as the lights of the car appeared she ran down the stairs and through the darkened entrance hall and got into the front of the car with the chauffeur. Only the night porters saw her go.

A minute or so later, Hinchliffe, Sinclair and the mechanic appeared in the door of the hotel. The mechanic turned to get into a second limousine which had now pulled in, while Hinchliffe and Sinclair opened the rear door of the first car and began to get in. Elsie Mackay signalled to Hinchliffe, moved up nearer the chauffeur, and tapped the vacant place beside her. Hinchliffe squeezed in next to her.

Sinclair, out of earshot in the back of the car, watched Elsie Mackay, with tears in her eyes, earnestly pleading with Hinchliffe. It was like watching a scene in a silent picture, heavily overplayed to compensate for the absence of sound. He had little doubt what she was asking.

When they got to the airfield, Elsie Mackay stayed in the car. It was broad daylight now. Hinchliffe called the second driver across and the two chauffeurs stood on guard, one on each side of the car. 'Keep anyone off who tries to speak to Miss Mackay,' said Hinchliffe. Even at this late stage Elsie Mackay believed her father capable of taking some action that might prevent her leaving. Like Lord Ullin's daughter before her, she would meet the raging of the skies, but not an angry father.

Hinchliffe and Sinclair walked across to the aeroplane. 'I've decided to take Miss Mackay after all,' said Hinchliffe. 'I'm sorry, but there won't be room for the three of us, so I shall have to leave you behind.' Sinclair nodded his acceptance of the inevitable. 'Miss Mackay asks that you say nothing of her departure. The relatives are not to know – not until we get there.' Sinclair agreed to say nothing and to disappear for a day or two. Meanwhile the mechanic was warming up the engines.

Two RAF officers were there to see them off, and presently an airman brought a typed weather forecast. Hinchliffe read it inscrutably and then walked across to the car. 'Everything is all ready now, Miss Mackay,' he was heard to say. 'The report seems good.' He held out the sheet of paper for her to read. 'You see, we have this east wind here, out in the Atlantic.

That will be a help.' They talked together for a moment, and then Elsie Mackay got out of the car.

The two fliers walked quickly across to the plane and stood for a moment together while photographs were taken. Then they said good-bye to Sinclair and the mechanic, and the two officers, neither of whom knew the aircraft's destination. 'Good luck and a safe journey,' said Sinclair. The next moment Elsie Mackay was waving to him through the lefthand cockpit window and Hinchliffe was giving the signal for 'chocks away'. With a roar of its engine the black-and-gold *Endeavour* moved off down the grass runway. Whatever the direction of the wind out in the Atlantic, a strong north-westerly breeze was filling out the windsock at Cranwell. Even so, the take-off run of the heavily-loaded Detroiter on the snow-covered grass runway lasted over a mile.

The departure from Cranwell was reported at once to the Air Ministry by signal, but no destination was mentioned. The plane was seen to set course in a westerly direction, and it was assumed that Hinchliffe was proceeding to an airfield in Ireland prior to a possible hop across the Atlantic. A stop in Ireland, indeed, may well have been in Hinchliffe's mind if weather or the behaviour of the aircraft warranted it. But the weather was good and *Endeavour* was behaving well. The only serious difficulties during tests had concerned the fuel supply and the new compass, and both were working smoothly.

After three hours' flying the Detroiter was seen near Waterford, in Southern Ireland, and although snow was falling again and the crew were getting only intermittent glimpses of land, the weather omens were otherwise favourable. Unknown to Hinchliffe, the United States Weather Bureau was reporting good flying conditions in the North Atlantic.

At 1.30 PM *Endeavour* was sighted over Mizen Head, in the extreme south-west corner of the Irish coast, setting course for Newfoundland. In five hours they had travelled 400 miles, giving a ground speed of 80 miles an hour. It was clear that they still had a headwind, but at their present rate of progress they would reach Newfoundland in another 23 hours – 28 hours flying in all. With a duration of 40 hours their safety margin was good, and with the prospect of a better ground speed as the fuel load lightened, and the expected change in the direction of the wind, they had a good chance to reach Philadelphia and win the long-distance prize.

Meanwhile the Press, learning of Hinchliffe's departure and getting evasive replies from Elsie Mackay's family and friends, soon drew the obvious conclusion. Further denials were pointless: the secret was out. In America, where there had been less need for secrecy, all the newspapers carried the story, with long descriptive passages about the flight.

In mid-Atlantic, however, the weather was deteriorating. The *Majestic* wirelessed that she was running into a south-westerly gale. The 16,000-ton liner *Republic*, also in mid-ocean, reported that she was barely able to maintain headway against westerly gales and mountainous seas. These were the sort of conditions that people had feared – conditions which, if they persisted, might force Hinchliffe to turn back. Yet on the far side of the Atlantic the weather was calm. The Newfoundland sealing fleet were cruising 150 miles out from the coast in good visibility and the sealing men were keeping a look-out for the *Endeavour*. They saw nothing, but this caused little alarm. Probably Hinchliffe was by-passing St John's and going all out for Philadelphia. But later that day, when there was still no news on either side of the Atlantic, anxiety grew.

By Thursday, 15th March, it was clear that *Endeavour* could no longer be airborne. There still seemed a chance that the plane had completed the crossing but that the two fliers had been blown off their course and were lost on the desolate, icebound coastline of Newfoundland or Labrador. There were many reports that the Detroiter had been seen in these areas, but all proved false on investigation.

As day after day went by with still no firm news, hopes for the survival of the two fliers faded. Encouraging rumours only cast relatives and friends into deeper despair when they were exposed as unfounded. At last only one possibility remained. Like so many would-be Transatlantic fliers before them, Ray Hinchliffe and Elsie Mackay had disappeared without trace.

There followed a train of events that proved almost as intriguing as the flight itself. A month after the tragedy, Hinchliffe's widow received a message supposed to have come from her husband through a spiritualist medium. She also received a corroborative letter from Sir Arthur Conan Doyle, author of the 'Sherlock Holmes' stories and a convinced spiritualist. Keeping an open mind, she attended a seance and received a detailed description of the flight, purporting to come direct from her husband.[1]

'After passing over Mizen Head,' said the message, 'we steered west-north-west for 850 miles. The weather was good but cloudy, and there was a little fog. This was between 2 PM and 10 PM and our speed was 80-90 mph.

'At 10 PM we began to encounter bad weather, but our spirits were high and we were still going well. We increased our speed to 100 mph up to midnight, and went in a more northerly direction.

'We had gone about 900 miles before we ran into the really bad weather.

'We ran right into the teeth of a terrific gale – wind, sleet, and rain. The wind broke the left strut and ripped the fabric. I saw that further headway was impossible against the gale, and I deliberately altered my course to the south, hoping to fly out of the gale and reach the Azores.

'We flew south from midnight to 3 AM, gradually coming down lower. At 1 AM, however, I knew that we were beaten. The compass had gone wrong and one of the plugs had oiled up.

'I carried on as best I could for two more hours, and at 3 AM landed in water one mile north of the Azores. We did not crash.

'I took a last drink from my flask and set out to swim for the shore. I swam for twenty minutes but the currents were too strong, and I became unconscious and finally drowned.

'Miss Mackay's end was peaceful. She was drowned in the machine while unconscious.'

1 *The Return of Captain Hinchliffe*, by Emilie Hinchliffe (Psychic Press).

The circumstantial detail of the flight, times, speeds and weather conditions cannot be faulted. Wherever they are susceptible of checking they are correct. Hinchliffe had always intended to turn north for Greenland if he ran into trouble, but he would instinctively turn away from so terrible a storm. At first sight it might seem that his most likely reaction would have been to turn back; but if he was half-way across the Atlantic, he might feel that there was almost as much hazard in turning back as in going on. There is no doubt that, at the time when he must have run into the worst of the weather, the Azores represented the nearest land.

The suggestion that Hinchliffe had tried to reach the Azores was unique; it had not been put forward as a possibility by any aviation authority up to that time. It ran counter to Hinchliffe's declared intention, yet it fitted the facts. But doubt was cast on the authenticity of this and other similar messages when, at the end of that year, an undercarriage wheel which was later indentified by serial numbers as belonging without any doubt to Hinchliffe's Detroiter monoplane was washed up on the coast of County Donegal. This suggested very strongly that Hinchliffe had been forced down somewhere on the direct route between Ireland and Newfoundland, and that very probably he had in fact turned back.

If the story put forward from occult sources was true, if the plane had come down so close to the Azores, how was it that wreckage had been found in Ireland and none in the Azores? Yet a study of the prevailing currents in the Atlantic showed that the incident was not wholly incompatible with the story said to have come from Hinchliffe. In a period of nine months the North Atlantic Drift could just feasibly have carried the floating wheel north-eastwards to the northern Irish coast.

But for the spirit message, though, one would certainly conclude that Hinchliffe turned back and got almost within sight of Ireland again before he was forced down. One might even suppose that he may have reached the coast and attempted a landing on some deserted, rocky shore.

After the tragedy it was discovered that the promised insurance cover, presumably through an oversight, had not been taken out. Lord Inchcape made good the deficiency. The residue of Elsie Mackay's estate after death duties – over half a million – went to the nation: the family did not wish to benefit under her will.

Exactly a month after the flight the German Junkers aircraft *Bremen*, flown by two Germans and an Irishman, successfully completed the first east-west Atlantic crossing, landing on an island in the Belle Isle Strait, Labrador after being blown off course. Three months later Amelia Earhart (as a passenger) became the first woman to fly the Atlantic.

At a celebration lunch for Miss Earhart in London on 25th June 1928, ten seconds' silence was observed in tribute to the three women who had failed where she had succeeded and died in the attempt – Princess Loewenstein-Wertheim, the American Mrs F. W. Grayson, and Elsie Mackay.

In 1936, when Emilie Hinchliffe married again, the bride was given away by an old friend of the family. His name was Gordon Sinclair.

8

A MISTAKE ON A MAP

CAPTAIN K. D. PARMENTIER, the 44-year-old chief pilot of KLM, walked into the offices of the Royal Dutch Meteorological Institute at Schipol Airport, Amsterdam, to be briefed about the weather. In about an hour's time, at half-past eight that evening, 20 October 1948, he was due to take off for Prestwick on the first leg of yet another Constellation flight to New York.

Dirk Parmentier had been flying Dutch airliners on the world routes for almost exactly 20 years. Born in Amsterdam in 1904, he had chosen aviation as a career when, as a boy of 16, he had joined Fokker's in 1920. Later he joined the Dutch Air Force, qualified as a pilot, and in 1929 transferred to KLM. He became one of the handful of pilots who, in the pioneering thirties, opened up the 10,000-mile air route to the Far East, from Amsterdam to Batavia. In 1934, with co-pilot J. J. Moll, he formed the KLM team for the England-Australia air race, won by C. W. A. Scott and Tom Campbell-Black. Parmentier and Moll, flying their Douglas DC2 (fore-runner of the Dakota) to a passenger schedule, outstripped all other entrants and finished only a few hours behind the winners. They won the handicap, and might have won the speed race too had not Parmentier, as always, put the safety of his passengers first.

The England-Australia air race established Parmentier in Holland as a national hero. It made no difference to his shy, modest manner and quiet dedication to airline flying. He was an experienced test pilot, too, his skill, approachability and readiness to seek advice earning him the task of test-flying all new KLM aircraft. Small in stature, he was not an impressive man at first sight, but unlike most national heroes he remained unspoilt, liked and admired by everyone.

In 1940, when Holland was overrun, Parmentier had taken off from Schipol Airport a short head in front of the advancing Germans, carrying a load of important Dutch passengers to Britain. For the rest of the war he had been chief pilot of the KLM crews who operated the regular schedules between Lisbon and the UK. It had been Captain Parmentier whose plane had been so fiercely attacked six weeks before the last fatal flight of Leslie Howard. How he escaped in broad daylight from the six Ju 88s which surrounded him became one of the great flying stories of the war, a superb piece of airmanship which saved the lives of passengers and crew.

The routine schedules of civil flying might not hold such excitement, but they were not without their moments of danger. In 20 years of flying, totalling nearly 16,000 hours in the air, Dirk Parmentier had proved himself to be one of the half-dozen great civil pilots, one of the men who were talked about wherever aircraft were flown. Now he was setting out on another Transatlantic service for KLM.

The weather over thc British Isles was poor; but when Parmentier called on the forecasters at Schipol he found them not unduly pessimistic. They warned him to expect intermittent rain, but they forecast that by the time he got to Prestwick the cloud there would be of less density than that being currently reported, indicating a general improvement during the two-hour flight from Amsterdam. The forecasters, however, were wrong; in point of fact the weather at Prestwick was deteriorating.

The Constellation, named the *Nijmegen*, was well equipped for sudden changes in the weather. Parmentier would be in touch with the ground by radio throughout the flight. Shannon, on the west coast of Ireland, had been nominated as the alternative jumping-off point for the Atlantic, and if the weather was bad at Prestwick Parmentier would fly on to Shannon. As a further safety measure, the *Nijmegen* carried enough fuel to fly to Prestwick, divert to Shannon, and then fly back to Amsterdam if necessary.

Walking from the meteorological office to the flight operations room, Captain Parmentier learned that there would be some delay to his schedule. Instead of flying the Atlantic direct from Prestwick, his routeing was to include a landing in Iceland. A consignment of freight for Iceland was now being loaded; it wouldn't take long, but the plane wouldn't be ready for an 8.30 take-off.

Eventually the *Nijmegen*, identification letters PH-TEN, left Schipol at eleven minutes past nine. Five minutes earlier the Prestwick wireless station had broadcast a weather forecast which indicated that in two hours' time it would be drizzling over the airfield and the cloud-base would be virtually solid at 600 feet. Had Captain Parmentier got away on time his wireless operator would have picked up the message.

It was vitally important to Captain Parmentier to know the precise amount and height of the cloud in the Prestwick area. His information from Schipol was that there was a strong cross-wind of about 20 miles per hour blowing at right-angles to the main Prestwick runway. The alternative runway was only 1,800 yards long and had a downward slope into wind, greatly reducing its effective length when it was wet. In any case the radar approach landing system was set up on the main runway only. In conditions of low cloud, and when the main runway was out of use through a strong cross-wind, KLM pilots were forbidden to attempt a landing at Prestwick. Dirk Parmentier, as chief pilot, had drafted these instructions himself.

There were other reasons why landing on the shorter runway in bad weather could be dangerous, especially at night. Inland of this runway the ground sloped upwards to a height, five miles east of the airfield, of over 400 feet. Three miles to the north-east the tops of the wireless masts, located on a hill, rose to nearly 600 feet. And a line of high-tension cables, running from north to south about three miles inland, created a lethal barrier east of the airport at 450 feet. Thus the greatest caution had to be exercised when circling inland. All this, of course, was well known to Captain Parmentier, who had flown in and out of Prestwick many times and who had written a full description of the terrain into the KLM route manual, with suitable warnings about the circling dangers east of the airfield.

After taking off from Schipol, Captain Parmentier headed north-west over the North Sea at 9,000 feet, crossing the English coast near Flamborough Head. He was carrying 30 passengers, including 11 Dutchmen, 6 Germans and a single Britisher. All were bound for New York.

The *Nijmegen* crossed Northern England obliquely to Carlisle and then began the traverse of Southern Scotland. Routine weather reports – as opposed to forecasts – were being broadcast by the Prestwick ground station, but none of them gave the cloud base as less than 700 feet. Shortly before eleven o'clock that night the Constellation came within R/T range of Prestwick, and Parmentier was advised by Prestwick ground station to call Prestwick Approach Control by voice radio. The forecast about the drizzle and the deteriorating cloud-base had still not been passed to him.

First Officer Kevin O'Brien, sitting in the right-hand seat of the front cockpit next to Parmentier, turned his control box to transmit and flipped on his microphone switch. A wartime RAF pilot, O'Brien had been with KLM for only four months but in that time he had sat in the co-pilot's seat for 16 landings at Prestwick and knew the airfield well. Now his breathy Irish intonation broke silence on the R/T.

'Prestwick Approach Control from Tare Easy Nan. Do you read me? Over.'

'Tare Easy Nan from Prestwick Approach Control. Loud and clear. Over.'

'Is GCA serviceable? What runway are they using?'

GCA – ground-controlled approach – was the radar system under which Captain Parmentier intended making his approach to land. As its name implied, it was a system in which the pilot relied for his information on verbal instructions from the ground.

'GCA is set up on Runway 32. The runway in use is Runway 26. GCA can give you overshoot on Runway 32.'

Confusing as this sounded it was clearly understood by

Dirk Parmentier. It meant that the approach system was set up on the main runway, the one with the cross-wind. He had expected this. To utilize the system he would have to make his blind approach through cloud down towards the main runway under GCA control. As soon as he broke cloud and could see the ground he would overshoot the main runway and execute a left-handed loop turn that would bring him downwind and alongside Runway 26, the shorter runway. He would then have to fly the length of this runway and a little beyond before turning back into wind for his final approach. It sounded complicated but once he was in visual contact it was quite a simple manoeuvre.

Success, of course, depended on breaking cloud somewhere over the main runway, which with a cloud-base of 700 feet he could expect to do. He would then ask for the main runway lights to be switched off and the smaller runway lights to be illuminated. GCA would have got him down safely under cloud and above the airfield, and the rest would be easy. In Parmentier's opinion this was far better than risking a landing with a full load on the main runway in a high cross-wind. He indicated his intention to O'Brien, who called Control.

'We will do GCA and overshoot on Runway 32, and will land on Runway 26.'

'Roger.'

At six minutes past eleven the coded radio broadcast announced that visibility and cloud ceiling were both deteriorating. But Parmentier and his crew were now in voice contact and were not using morse communication. No message about the worsening of the weather was passed to them by voice radio.

Captain Parmentier was now proceeding on two assumptions, both of which were false. One was that the cloud-base was no lower than 700 feet. The other was that the cross-wind on the main runway, which he had been told before take-off was 20 miles per hour, was unchanged. In fact, the cross-wind had dropped. It was now only 14 miles an hour – just inside the KLM limits for landing a fully loaded Constellation on the main runway. The second of these misapprehensions –

but not the first – was corrected in Parmentier's mind at sixteen minutes past eleven by a voice message from the ground.

'Tare Easy Nan from GCA Controller. I would advise you that the wind is south-west, 12-15 miles per hour. You may land on Runway 32, and if you find the wind too strong you may land on Runway 26.'

This was good news for Captain Parmentier. He had always urged his pilots to accept a cross-wind of up to 15 miles per hour on the main runway here at Prestwick rather than use the shorter one. Unless he encountered high gusts on his final approach, there would be no need for the complicated overshoot procedure at all. He signified his agreement to O'Brien, who called the controller.

'Roger. We will attempt to land on Runway 32.'

Following instructions from the controller, Captain Parmentier reached a position 20 miles north-east of Prestwick at 4,000 feet, losing height at 500 feet a minute. He throttled back to circuit speed and carried out his cockpit check for overshoot and landing. Two minutes later, on instructions from the controller, he turned south, still about twelve miles inland, and continued to fly a southerly course until he reached Cumnock, eleven miles south-east of the airfield. He had now let down to 2,000 feet.

'Turn right on to heading 270.'

'Roger.'

'If you lose radio contact on this approach, fly at 4,000 feet and call Prestwick Approach Control.' This was a safety measure, ensuring that the pilot climbed clear of trouble if control was lost.

'Roger.'

After two minutes on the new heading, in which he reduced height to 1,500 feet, Parmentier was flying on instruments through dense cloud. He was eight miles from the airfield, approaching from the south-east.

'Turn right, heading 305.'

The controller was lining him up on the main runway. He was maintaining his height at 1,500 feet, still in cloud.

'Check undercarriage and flaps for landing. I am advising

you of a strong cross-wind from your left on this runway. Do not acknowledge further instructions.'

Kevin O'Brien operated a switch in the cockpit, and back in the cabin an illuminated sign flashed on for the fastening of seat-belts. Elsie Fey, the 25-year-old air hostess, checked that all the passengers were properly strapped in.

The Constellation seemed strangely quiet as it ploughed its way through the darkness and cloud. This was partly due to the throttling back of the engines, partly to deafness caused by the change in air pressure. The thick wall of cloud and fog itself seemed to deaden sound.

'Seven miles. Turn left five degrees on to heading three-zero-zero. You're coming along nicely now ... 6½ miles.'

In the GCA van adjacent to the main runway, the controller watched the progress of the *Nijmegen* on his precision screen, prompted at intervals by the man at his side, the elevation tracker, who was keeping a check on the aircraft's height. Meanwhile Parmentier was maintaining his heading and losing height steadily.

'Heading three-zero-zero, still OK. You're coming on to the glide path now. 5½ miles.'

The controller's voice was calm and reassuring, no more than a whisper in the ear, yet of startling clarity.

'Heading 295. Three miles to go.'

The wind that was blowing across the main runway was still crabbing them to starboard. Captain Parmentier pressed the control column to the left, watching the gyro compass swivel through three degrees. Although the strength of the surface wind was only 15 miles per hour, Parmentier judged that it must be considerably stronger above ground level, enough to make the drift on his final approach difficult to control. He decided after all not to attempt a landing on the main runway, but to carry out the overshoot procedure and land on the shorter runway.

'Hello controller. I shall overshoot on this approach and land on Runway 26.'

'Roger. The surface wind is 15 miles per hour. Give me a check call when you are downwind for Runway 26.'

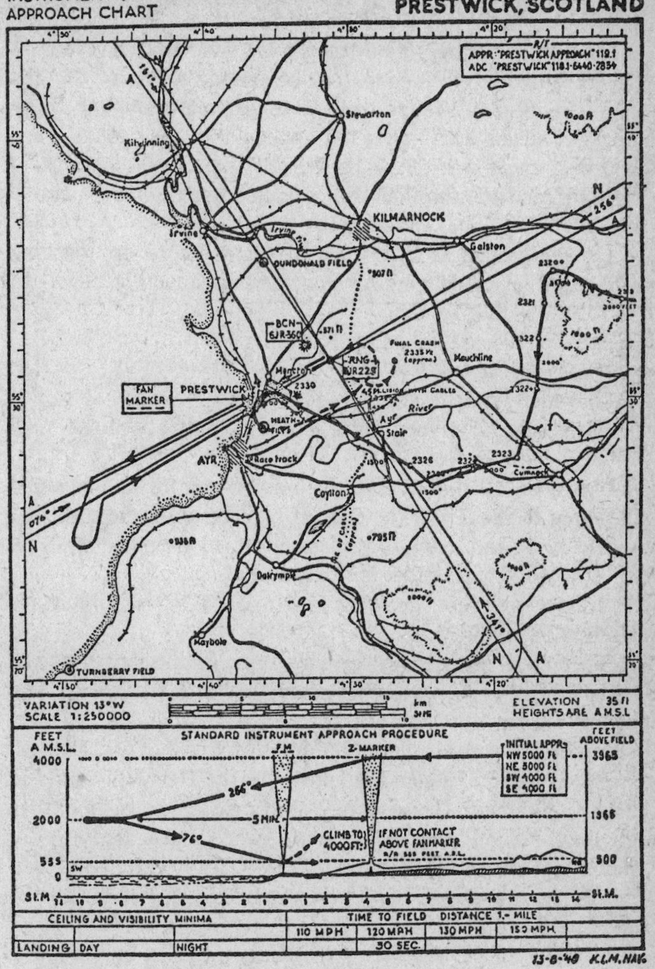

5. The final approach path of the *Nijmegen*, showing the point of collision and the final crash

They were down to 700 feet now, the height of the cloud base – according to the reports received from the ground. And in deceptive corroboration, they were suddenly conscious of speed as swirls of cloud parted before the thrust of the aircraft's snout. They found themselves peeping through the dark opaque curtain as large rents appeared in the cloud. Then ahead of them was the warm friendly pattern of the airfield lighting, and the familiar trick of convergence of the parallel runway flares. They were below the cloud.

On the ground at Prestwick, the lights of the approaching Constellation could now be seen, gliding smoothly down towards the main runway.

'Are you visual?' asked the controller.

'Yes, I am visual.'

'Turn left three degrees. 1½ miles to go.'

Parmentier continued until he was directly above the main runway at 200 feet. Then he opened his throttles gently, carrying on to the end of the runway before banking and climbing to the left at the start of his loop turn, the turn that would bring him back towards the airfield from the west, on his downwind leg for the shorter runway.

'I have overshot. I am going to do a visual circuit for landing on Runway 26.'

There was no need to time his turn, or the length of the downwind leg. He was in visual contact with the ground and would remain so; there were patches of drizzle, but visibility was more than two miles. The lights of Runway 26 had been switched on and he could see them clearly. He swept round in a fairly tight turn which barely overlapped the coastline, then climbed to 450 feet and selected wheels and flaps down, flattening out for the downwind leg. The Constellation ballooned slightly as the flaps were extended. Almost at once he ran into cloud.

The cloud base had been consistently given, in those reports received by Parmentier and his crew, as 700 feet, and this had been confirmed by their own experience on the let-down to the main runway. They had had no warning of any deterioration. There was nothing unusual, they knew, in meeting an isolated patch of cloud several hundred feet below the reported cloud base. In a moment they would have run through it.

Captain Parmentier did not know that two large airliners of Scandinavian Airlines, whose pilots had received the forecast of the lowering cloud base, had turned back for Copenhagen. He kept his eye on his instruments while Kevin O'Brien stared across to port, waiting for the runway lights to reappear. It could be no more than a moment. The airfield flashing beacon was at the bottom of Runway 26, and a funnel of orange lights pointed the way in to the runway. Both had been clearly visible a few seconds ago.

The weather forecast that had failed to reach Dirk Parmentier was now being corroborated in fact. The occasional scattered cloud that had loitered below the main cloud base was now joining hands. As low as 300 feet, the area to the south and east of the airfield was almost completely blanketed by cloud. Had Parmentier known this he would never have attempted a landing. All his instructions strictly forbade a landing on Runway 26 when the cloud base was below 700 feet.

The deterioration was well known to the approach control staff at Prestwick, but through faulty procedures they had failed to inform the pilot. Parmentier continued on his downwind leg, confident that he must break cloud shortly. With the weather information in his possession it must be so. But the cloud remained solid.

Both pilot and co-pilot were now peering across to port for a glimpse of the runway lights. Because of the strong wind at their backs, they were in fact already past the airfield and flying inland. Two miles distant, right on their present track, and at their exact height, lay the deadly obstruction of the high-tension cables. Two miles distant. In terms of flying time in the *Nijmegen*, less than a minute.

It seems unbelievable that Dirk Parmentier could have forgotten the existence of these cables, or their height. Yet for the moment he did not change course or altitude.

On his knee Kevin O'Brien was balancing a chart of the airfield on which approach lanes and obstructions were shown.

It was an official chart printed and issued by KLM. But by some inconceivable error, the spot height shown at the highest point of the high tension cables, instead of being three figures – 450 – was 45 feet. More incredible still, a graph showing the ground contours undulated to heights of no more than 250 feet, proving that the spot height of 45 feet shown on the chart was no mere isolated error. Anyone circling to the east of Prestwick airfield on the basis of this chart was certain sooner or later to crash into high ground.

Permentier and O'Brien were still staring through the port window for a glimpse of the airfield lighting. But as they neared the end of their downwind leg the cloud and fog intensified. Because of their expectation that they would be able to fly the circuit visually, they had not timed the leg exactly. Now they were uncertain of their position. It seemed – looked at in retrospect – a situation in which they ought to have broken off, climbed to safety, called Prestwick Approach Control, and started all over again. But the *Nijmegen* held its course. Still at 450 feet, they were less than half a minute from the high-tension cables.

Surely within seconds they would take avoiding action. Surely they could not be wholly unaware of their danger. Captain Parmentier had himself phrased the warnings about this high ground inland from Prestwick. Both he and his co-pilot were thoroughly familiar with the airfield. Could they possibly be deceived by an erroneous chart?

It could be that the powerful tail wind had blown them much further inland than they imagined. Or were they, perhaps, pre-occupied with some other emergency, such as a failing engine? Whatever the situation was, the two men evidently decided to call Prestwick Approach Control for re-orientation. The voice of Kevin O'Brien came up loud and clear.

'Prestwick Approach Control from Tare Easy Nan. Do you read me? Over.'

'Tare Easy Nan from Prestwick Approach Control. Five by five. Over.'

When O'Brien made his call, the *Nijmegen*, still flying completely blind, was only seconds away from a collision that even now was not inevitable. A tug on the control column and the Constellation could still climb clear.

Stretched between steel pylons 1,000 feet apart lay the four cables of the national grid system. The pylons themselves were over 450 feet high. But the *Nijmegen* was flying directly between two of the pylons, and the cables sagged in the middle. There was still just a chance that the aircraft might clear them.

The topmost cable was the earth cable. If the aircraft hit this, it might escape destruction. But the three lower cables were phase conductors constituting the main grid line from Kilmarnock to South Scotland. They carried 132,000 volts.

There was no time for O'Brien to follow up his preliminary call to Prestwick Approach Control. In the next instant, all the lights on the airfield flickered momentarily. Then there was a clipped, fading call from O'Brien.

'We have hit something ...'

It was the three live cables that the *Nijmegen* had struck.

The voice of Dirk Parmentier was then heard for the first time, calmly giving instructions to O'Brien. The aircraft's transmitter was still switched on.

'Operate fire control.'

Captain Parmentier was faced with the two most mortal dangers known to airmen – fire, and damaged controls. O'Brien operated the internal fire extinguishers in the engines, but they had little or no effect. The high voltages had fused all insulation, petrol in the tanks had ignited, and fire had broken out in the cabin. The passengers were hurriedly unstrapping themselves to get away from the fire.

'We are climbing.'

Parmentier still hoped to get his crippled plane down safely. Opening the throttles, he was trying to climb away from the high ground and at the same time turn back towards the airfield. He was still flying on instruments in dense cloud.

The flickering of the lights on the airfield was not at first connected with the circuit of the Constellation. But a minute or so later a 'blip' was seen on the precision tube in the GCA van, at a point about five miles east-north-east of the airport, near the village of Tarbolton. It was reported to Prestwick Approach Control, who called the aircraft.

'What is your position?'

There was no reply. Yet the Constellation was still airborne. It had flown into a left-hand turn as Parmentier strove to regain the airfield, but this turn had become uncontrollable as the airliner described a complete circle to port. The villagers of Tarbolton, drawn from their homes by the roar of engines racing at full power, glimpsed the low-flying aircraft careering across the countryside in a torch of flame.

'Have you any idea where we are?'

It was the last despairing call from the *Nijmegen*. Three minutes after the collision with the electric cables, still locked in its immutable turn to port, without having lost or gained height, and with the voice of the ground controller calling it in vain, the Constellation crashed into high ground five miles east-north-east of the airfield at lonely Auchinweet Farm.

The first man to telephone news of the crash was the village policeman at Tarbolton. But because of misunderstandings and confusion over responsibility for rescue services, and persistent fog, medical aid did not reach the scene of the crash until one hour forty minutes after the accident. By that time only six people were alive. All six died within 24 hours without throwing any further light on the cause of the disaster.

The main problem for the subsequent court of enquiry was to establish why the aircraft had flown towards the high ground east of the airport at such a dangerously low altitude. They were unable to find any one factor chiefly responsible, but they concluded that what they called a 'pyramid of circumstances' had combined to produce the tragedy. First there was the complete absence, in the radio messages and conversations between aircraft and airport, of any reference to a deterioration in the weather. Second was the failure of the crew to employ a timing procedure when flying downwind of the south-west runway – Runway 26. Third was the false and misleading character of the chart in use. It transpired that these charts had been copied from American Air Force maps, which must also have been faulty. The court expressed astonishment that such charts should be based on a foreign authority when detailed and accurate Ordnance Survey maps were available.

It remains a mystery why Captain Parmentier and his First Officer, well acquainted with Prestwick and its particular dangers, should have trusted their charts when it would seem that a moment's consideration must have shown them to be erroneous. But can it be stated categorically that they did follow their charts? Is there any other possible explanation of their low-altitude flying in this dangerous area? The state of No. 2 engine after the crash suggested that it might have been feathered at the moment of impact with the cables, and engine failure, coupled with unexpected low cloud, could certainly account for a measure of pre-occupation on the part of pilot and co-pilot. But would this be sufficient to make them careless of a known and feared hazard?

The chart was meant to present pictorially, for quick and easy reference in the air, the information contained in the more verbose flight instructions compiled by Captain Parmentier as chief pilot of KLM. It is probable that Parmentier and O'Brien had used the chart many times when landing at Prestwick, and there is evidence that it was actually in use in the cockpit on this occasion. If the two men had noticed the error in the spot height they would obviously have corrected the chart; so whether the chart had survived the fire and was still legible was a question of vital significance to the conclusions of the court.

'The actual chart,' says the report of the enquiry, 'was recovered in a half burnt and charred condition.

'It is a poignant circumstance that while about one-third of the chart was completely burnt, the charring stopped within one-eighth of an inch from the place where the false marking of "45 feet" appears.'

The chart was uncorrected.

Just as the old barnstorming pilots flew by the seat of their pants, so the civil pilot relies on his instruments. And as the private pilot might rely on his memory for the configuration of a particular airfield, so an airline pilot like Dirk Parmentier, using many airports on the world's airways, and conscious of the fallibility of human memory, might train himself to rely on his charts.

This must be the explanation why Captain Parmentier, one of the safest and finest civil pilots of all time, flew into high ground on which he himself had sounded grave warnings to all his pilots only a few weeks earlier.

9

THE WRONG DOOR

THE man who got out of the polished black limousine outside the grey stone airport buildings at Croydon had the powerful shoulders and square, thick-set physique of a cruiserweight boxer. He was in fact – or had been until recently – one of the fittest men for his age in Europe. He was also – or had been until recently – one of the richest. But within less than two hours of his arrival at Croydon Airport – the date was 4th July 1928 – the state of his health and finances would be the subject of international speculation and gossip. For Captain Alfred Loewenstein [1] was about to embark on one of the most dramatic flights of the century.

Within the limits of human fallibility, evidence which seems incontrovertible can be produced to show that, a few minutes after his arrival at Croydon, Loewenstein boarded his plane with four members of his staff plus pilot and co-pilot and took off on a cross-Channel flight. Similarly incontrovertible evidence can be produced to show that when the plane landed in France less than an hour later he was not in it. Short of some almost unimaginable collusion amongst his staff in smuggling him out of the plane, either just before take-off or immediately after landing, Loewenstein had disappeared in mid-Channel. But this is a story in which nothing can be taken for granted. Every move that Loewenstein made before the flight, every word he uttered, may be of significance. So, too, may be the circumstances of his past life.

Even for the roaring twenties, the pace and opulence of Alfred Loewenstein's life were something to be marvelled at. He was one of the first men to attempt to live at the speed and

[1] No relation to Princess Loewenstein-Wertheim of story No. 3.

scale made possible by the inventions of the twentieth century. Most spectacular was his fleet of aeroplanes, piloted by seconded captains of Imperial Airways, standing by to transport staff and business associates – and even sometimes nothing more than a letter or a verbal message – from one European city to another. This gave to his operations two qualities which he regarded as vital – speed, and secrecy.

Loewenstein was the son of a prosperous Belgian banker. Accustomed to luxury, he was left penniless at 19 when the family business failed. But the young Loewenstein was a financial genius, a gambler and speculator gifted with extraordinary prescience in choosing his investments besides being adept at manipulating the market. Long before the First World War he had built up a fortune, and after the war, with inflation bedevilling all European currencies, he made enormous gains. Indeed his detractors accused him of battening on ruined and half-starved Europe in unscrupulous deals, buying on credit and relying on currency depreciation to ensure enormous profits when settling day came.

For most of his life Loewenstein had preferred to be the power behind the scenes, but in 1926 he gained fame and notoriety overnight by his highly publicized offer to form a syndicate which would lend £10 million sterling to the Belgian Government, free of interest, in an attempt to stabilize the franc. The news had hardly been digested when he followed it with an offer to lend £30 million to the French Government for the same purpose, though in this case he asked for nominal interest of 2 per cent. Loewenstein would ask in return for the delivery of an equivalent sum in Belgian and French francs at current rates of exchange, which he would use to buy shares in Canadian companies operating in France and Belgium. By means of this and other purchases a steady stream of currencies possessing a gold value would enter the French and Belgian Treasuries.

Loewenstein imposed two conditions: that during the loan period the governments concerned would not borrow except through his syndicate; and that they would accept a plan to be proposed by his syndicate for currency stabilization. These conditions seemed difficult to reconcile with the sovereign powers of governments, and in any case other disadvantages were foreseen. The net result, it was felt, might be that the governments would soon get back their paper francs and inflation would result, while the market value of the securities bought by Loewenstein would tend to rise. At the end of the loan period Loewenstein would return his millions of devalued francs and get back his sterling. It would be good business for Loewenstein, but for France and Belgium the last state would be worse than the first.

The offer had one positive result, however – it established Loewenstein in the public mind. Very probably this was what he wanted. The public became accustomed to reading about this large, awkward, quietly-dressed superman who lived like a prince out of a Ruritanian extravaganza and made and spent millions. Forty villas at Biarritz were hired for his guests and entourage when the franc stabilizing deal was being put forward, and he hired a whole floor of the Ritz or Claridge's when he came to London. In addition he had a large private estate near Melton Mowbray and a palatial home in Brussels.

Once the limelight had picked him out, Loewenstein found it difficult to avoid. A man of rapid changes of temper, volatile and impulsive, he made bitter enemies as well as loyal friends. Hot-tempered and even violent on occasion, his rages were said to be quite abnormal. In 1926 he was fined at the Bayonne court for two assaults, one on the doorman of the casino at Biarritz when he was refused admission because his companion was not in evening dress, the other when he became so enraged at the behaviour of one of his private secretaries that he attacked him and struck him several times. Such lack of physical control suggested, perhaps, a mental unbalance. Or it may have been that his overworked mind needed an outlet. A man of stupendous energy, he made a fetish of physical culture and carried on his staff his own boxing, tennis and golf professionals.

Wherever he went he was news. He attended the opening meeting of the Quorn by flying up from Croydon. He competed in the horse-jumping show at Olympia, driving his horse almost as mercilessly as he drove himself. He became well-known as a racehorse owner, and he bought the favourite, Easter Hero, in a vain attempt to win the 1927 Grand National. His reputation of international financier became inextricably woven with that of international sportsman. The public gambled on the shares of his companies. His interests were wide and included mining and hydro-electric power in many countries, but his dominant aim and ambition was to create and control a world artificial-silk combine.

At first he seemed to be succeeding. He bought his way into British Celanese, gaining control of the Board, but he lost it when the Dreyfus brothers, founders of the firm, bought up every share they could lay their hands on until they were able to oust him. But he would not accept defeat. He tried again and again to manoeuvre his way back into a controlling position, threatening all kinds of action against the brothers Dreyfus, who stood firm.

Although he was a multi-millionaire, most of his money was spread in share transactions where fortunes could be made or lost in a single day. With the first hint of a slump on Wall Street he lost millions. Some of his associates were ruined. He declared that he had finished speculating and was looking now for stability. He was in his fiftieth year, and there were rumours of a breakdown in health. Yet he continued his personal vendetta with the Dreyfus brothers over British Celanese. Eventually, in May 1928, Henry Dreyfus was obliged to take a full double-column of newspaper space to deny that his company had anything to do with Loewenstein. 'Approaches have been made in this direction,' said the statement, 'but since relations were broken off these approaches have been consistently ignored or refused and in view of past experience will be ignored or refused in the future. As a consequence of this attitude on our part numerous attacks have been made by Mr Alfred Loewenstein both on myself, on my Group and on the Celanese Companies. These attacks have been treated with contempt . . .'

The last part of the statement took the form of an ultimatum to Loewenstein to lay off. 'Whilst my Group and myself and the Companies connected with us have deliberately abstained, in spite of provocation, from taking action, we are of course watching very carefully in the interests of our business the activities of Mr Alfred Loewenstein . . . so that, should the occasion arise, we should be in a position to prove any facts required.'

Just what it was that underlay this implied threat is not clear. But the night before the flight a private detective appointed by Loewenstein had discovered the origin of an anonymous and libellous 20-page document which comprised a vicious and detailed attack on Loewenstein. It accused him of being a parasite who created nothing, whose interests from start to finish were bound by the market and ways of manipulating it. It alleged that he had abused his position in the International Holdings Company – his main company – by turning his managership into an affair of personal gain, without regard to the welfare of the company. It forecast that he planned to start a 'bear' attack on the shares of British Celanese in an attempt to reduce the value of the Dreyfus holdings and force them out. For good measure it accused him of staging a jewel robbery at his Biarritz villa to collect £200,000-worth of insurance. And finally it arraigned him for attracting millions of pounds from small investors for schemes which were not suitable for public investment at all. All these allegations were supported by a wealth of detail in the document, which the blackmailer threatened to circulate to every banker, stockbroker and finance house in Britain, Europe and America unless Loewenstein paid up. If a fraction of the accusations were true, Loewenstein was finished. But he himself was satisfied that he had categoric answers to all the allegations, and now he knew the author of the document he was ready to plan a counter-attack.

The blackmailer almost certainly knew he had been discovered. Here, then, was a man with a motive – perhaps even a motive for murder.

Loewenstein, then, was ready to deal with his blackmailer. What about Dreyfus? Just before he took off from Croydon he was asked by a reporter whether there was any possiblity of a truce with Dreyfus. His reply was vehement. 'Never! I am busy now, but when I come back from Brussels I shall have more time. Then I shall settle Dreyfus once and for all.'

This was the man who pushed his way through the swing doors into the main entrance hall of the airport building at Croydon, where he was met by the man who for the past nine months had been his co-pilot, navigator and mechanic, the 27-year-old Bob Little. It was a perfect July evening. The time was six o'clock. Apart from a slight trace of tiredness that any business man might betray after a hot summer day in the City, Loewenstein was his usual robust, affable self. He quizzed Bob Little in characteristic manner, his Continental accent adding urgency to everything he said.

'What sort of flight are we going to have, Mr Little? How long will it take? Are we going to be late? What height are we going to fly?'

'The weather's perfect, Captain.' Everyone who knew Loewenstein addressed him as captain; he had held the rank in the Belgian Army in the First World War and was proud to retain it. 'We'll be in Brussels in about two hours. Height about 4,000 feet.'

Loewenstein looked relieved. He hated flying high. 'I've got a phone call to make. Then I'll be with you.' He walked across to the KLM office on the right of the main hall and asked to use the phone. The door of the office was open, and presently Loewenstein's voice could be heard discussing business matters with Sir Herbert Holt, the chairman of International Holdings.

'Will you write to me on that?' There was a pause. 'Good. How about coming over for the weekend? Can't you manage it? Anyway I'll be back in a week or ten days and we'll discuss it then. Good-bye, Herbert.'

As he came out of the office he walked into Captain R. H. McIntosh, a British Airways pilot who had previously flown for Loewenstein. The two men exchanged greetings. McIntosh thought Loewenstein was looking well and in the best of spirits. 'Donald Drew's waiting for you in the plane,' he said.

Drew was Loewenstein's personal pilot, tall and dark, with pointed features, seconded from Imperial Airways.

'Right. We're just off. So long, Mac.'

Followed by his entourage – Hodgson, the secretary, Baxter, the valet, and Mlle Bidalon and Miss Clark, shorthand-typists – Loewenstein passed through Customs and out on to the tiny square of tarmac in front of the airport buildings. Drew was waiting for him at the door of the plane. It was a three-motored Fokker VII, built in Amsterdam. The starboard and centre engines were ticking over, but the port engine had been left until the passengers were aboard. The door was on the port side.

There was a small strutted external step, which remained in position in flight. Loewenstein climbed up easily and into the rear of the plane. Immediately opposite, as he entered, was a wash-basin and mirror, just aft of a toilet door which opened towards the plane entrance. This was a dual-purpose door, hinged forward, which also closed on the cabin, shutting off the cabin from the rear of the plane in flight.

Loewenstein's was a specially fitted plane, and this dual-purpose door was a refinement. Its only use was to hide the lavatory seat while passengers were embarking and disembarking. In flight it was always closed on the cabin, leaving the whole of the toilet and wash-basin area clear.

Loewenstein stepped into the richly carpeted cabin and went forward to his armchair in the front on the left, facing forward, with a table in front of him. Diagonally opposite, on the starboard side facing aft, sat Arthur Hodgson, the personable secretary, also with a table in front of him. The arrangement was intended to allow the two men to face each other and converse, but this was rarely possible above the noise of the engines. Further back in the cabin Fred Baxter, the diminutive valet, sat on the port side behind Loewenstein, with the two typists on the right, facing each other, with a table between them. This was the lay-out inside the cabin as witnessed by Bob Little, who was the last to board the plane. He had stayed by the port engine while it was started, climbed the step, and finally turned the handle of the external door from the inside so that it could not spring open. There was a tiny lock in the handle, useful for securing the plane at night, or during the day when personal belongings had to be left inside after landing, but it was never locked in flight.

As Little walked through the centre of the cabin to the cockpit he passed Loewenstein – or a man whom he took to be Loewenstein – in the front left-hand seat. He didn't stare at the man for the sake of positive identification – he had no reason to – but it did not occur to him to doubt that it was Loewenstein. All the other passengers were safely on board. There was a moment's delay while Baxter got out of the plane to speak to someone on the tarmac; then he got back in and closed the door.

One of the girl secretaries, Ellen Clark, was waving to a man friend who was standing in front of the airport building. His name was Cox, and as he waved he recognized Loewenstein – or thought he did – looking out of the cabin window, sitting in his usual seat.

There were no runways at Croydon then. All Donald Drew had to do was to taxi 50 yards to the northern boundary of the airfield and turn into wind. Then the lush grass began to flatten as Drew opened the throttles and began his take-off run. Soon they were turning over Purley and heading for Sevenoaks, then Maidstone and Ashford, crossing the fruit-laden orchards of Kent. Donald Drew relaxed and was soon absorbed in a novel; Bob Little flew the plane from the right-hand seat. It was a beautiful evening, without a cloud in the sky, and the air was clear and almost still. At such a time, and with the confidence that came from having three motors, all running sweetly, it was a joy to be flying, a joy to be alive. Something of this contentment was present in the cabin, though there was not quite the same air of relaxation. It was more like an office than the inside of an aeroplane. Typewriters were clicking, documents were being passed from hand to hand, Hodgson was busy drafting letters and scribbling notes.

By the time he got to the coastline Little was outside the range of Croydon with his voice radio, but he called the airfield at Lympne. 'Hello, Lympne. York Ink calling. I am leaving the coast at South Foreland at 4,000 feet, heading for Dunkirk. My destination is Brussels.'

Drew did not go back into the cabin to see how his passengers were enjoying the flight. Loewenstein and his staff would be busy, and Loewenstein would not have welcomed the interruption. While Little kept the Fokker on course, Drew carried on with his novel. Little shot an occasional glance back into the cabin through the window of the cockpit door. Soon after they crossed the coast he saw Loewenstein with his head half out of the sliding window by his seat, apparently looking down at the Goodwins. Visibility was unlimited, and Little could see straight across the Channel to Cap Gris Nez, north as far as Harwich and south to Le Treport and Dieppe. No doubt Loewenstein, too, was enjoying the view.

There was, of course, another interpretation. Loewenstein had been suffering from overstrain. It was rumoured that his heart was affected. Four thousand feet was not high but the oxygen content would be lower than at ground level. He had taken off his coat, which was not unusual, also his collar and tie, which was. It seemed to Hodgson that he was trying to get more air.

Before they were half-way across the Channel, Loewenstein got up from his chair and walked down the cabin to the toilet, closing the door again on the cabin. Several minutes passed, but he did not reappear.

Bob Little was still flying the plane. Donald Drew was still absorbed in his novel. Hodgson went on writing his notes. The typewriters went on clicking. But Alfred Loewenstein was never seen alive again.

Nearly ten minutes had elapsed before anyone suspected that something might be the matter with their employer. Hodgson went back and questioned Baxter, and the valet got up and knocked on the door leading to the toilet compartment. There was no reply. He rapped again loudly and insistently, but there was still no response.

While Baxter stood back, pale with apprehension, Hodgson grasped the door handle and pushed. He stared into the compartment incredulously. It was empty. But the outside door of the fuselage was open, quivering very slightly in the slipstream. Loewenstein, it seemed, had fallen out.

Baxter collapsed against the arm of his seat, his teeth chattering with fright. The two girl secretaries, wide-eyed with horror, were on the verge of hysteria. Beads of sweat stood out on Hodgson's brow.

The impossibility of carrying on any sort of coherent conversation against the high engine noise heightened the stunning effect of the tragedy. Even Hodgson suffered momentarily from mental chaos and confusion. The truth stared him in the face, yet he could not believe it. He stood there like a man who has seen the whole sure edifice of his life crumble and vanish. Then, staring round the cabin as though in a moment his eye must alight on his missing employer, freeing his mind from its present fantasy of horror, he sleepwalked back to his seat. There his training came to his aid. Picking up his notebook, he scribbled three words on a blank sheet of paper and tore it from the pad, then went forward to the cockpit.

Putting his novel aside, Drew took the proffered slip of paper without looking round. He read it idly at first, without comprehension.

'The Captain's gone.'

Again the noise of the engines drowned coherent questioning and thought. Drew tapped Little's arm and showed him the message. Then he left his seat and went back into the cabin.

'Gone? What do you mean, gone?'

'He's not here.'

'Not here?'

'He's fallen out.'

The scene in the cabin had about it all the grotesqueness of nightmare. The two typists were slumped across their table, one on each side, their wide-brimmed hats hiding their faces but not their sobs. Baxter, pale as death, trembled as though with an ague. But with the aid of gesticulations and lip-reading Hodgson and Drew somehow made themselves understood.

It did not take Drew long to verify Hodgson's statement. Loewenstein had certainly disappeared. He hurried back to the cockpit, took over from Little, and began spiralling down. Straight ahead of him were the beaches between Gravelines and Dunkirk. He had almost completed the Channel crossing, and no one knew exactly when Loewenstein had fallen out. After a perfunctory search at low level which revealed nothing, he decided to land on the beach as quickly as possible and get some sort of story from Loewenstein's staff. That was the first essential – to establish what had happened and get their story sorted out. He brought the Fokker down over the pale flat beaches and landed safely. It was an unorthodox thing to do, but it was better than facing the probing questions of airport officials and Press reporters before they were ready for them.

At last, away from the din of the engines, it was possible to examine the plane and interrogate Loewenstein's staff. If Drew had any doubt whether Loewenstein had in fact boarded the plane – and he was ready to swear he had seen him do so – all suspicion was removed by the sight of his coat draped over the back of his seat, straddled by his collar and tie, thrown down untidily by. Little, too, was ready to swear that he had seen Loewenstein with his head half-out of the window, looking down at the Goodwins. Although both men realized that it was impossible to prove that a clever confidence trick of substitution had not been worked on them, both had no doubt in their own minds that Loewenstein had been aboard.

They were just as certain, too, that he had *not* been aboard when they landed. But doubt was later to be thrown on this point as well. Loewenstein, it was said, had somehow spirited himself off the plane before take-off, or hidden himself in some secret compartment aft of the lavatory and crept away on Dunkirk beach. His presumed object was to escape from some impending financial crisis by pretending to be dead. The second suspicion – that he had slipped away at Dunkirk – seemed more than feasible; Drew's impulsive act in landing on a lonely beach where no Customs or immigration officials were present was certainly open to misinterpretation. However, bearing in mind Drew's reputation and status, and the sworn evidence of staff and crew, neither theory deserved serious consideration and both were soon discredited. There remained three possible explanations – murder, accident, or (and it seemed by far the most probable) suicide. When, on the following morning, shares in the Loewenstein companies slumped dramatically and rumours grew of a financial crash, credence was given all too hastily to the theory of suicide. Such an explanation accorded with the popular view that big-time stock-market speculators who lived lives of luxury generally came to a sticky end.

The suicide theory was strongly supported by a series of tests made on similar planes, which showed that it was difficult, indeed almost impossible, to open the outside door more than an inch or two in flight – except by applying quite abnormal pressure. This seemed to dispose of the idea that the door could have been opened accidentally. But tests carried out by the Inspector of Accidents for the Air Ministry, made not on similar planes *but on the plane itself*, showed that it was in fact quite easy to open the outside door in flight, anyway up to about eighteen inches to two feet. So accident remained a plausible theory.

There remained the third possibility – murder. The blackmailer was the first suspect, and there may have been others who for their own reasons wished Loewenstein out of the way. But Loewenstein was a man of great physical strength; unless he were drugged he would be difficult to overpower. Murder was not impossible but it seemed unlikely.

Enquiries were complicated by the difficulty of defining responsibility. Loewenstein was a Belgian, he had been travelling in a Dutch plane that had been registered in England, the accident had occurred over the Channel, outside territorial waters, and the plane had landed in France. Small wonder that rumours multiplied and the true facts remained obscure. But when, 15 days later, Loewenstein's body was picked up by a trawler in mid-Channel, most of the wilder theories were discounted. The autopsy revealed no serious illness, no trace of a struggle, or of poison, or of drugs. Loewenstein had been alive when he hit the water, he had suffered multiple injuries, and the actual cause of death had been drowning.

The behaviour of the valet Baxter made him a possible

suspect; he had left the plane for a few moments just before take-off, and he acted oddly when Loewenstein disappeared. But there seems no reason to doubt that he was a trusted and trustworthy servant. Did he, perhaps, forget to close the outside door? This and his apparent neurotic tendencies could explain his subsequent suicide, four years later, while in the employ of Loewenstein's son.

Looking back over everything that was written at the time about the incident, two facts stand out. First, all those who didn't know Loewenstein, or his plane, took the cynical view – that Loewenstein had committed suicide. Second, all those who knew both the man and the plane were adamant that the notion of suicide was inconsistent with Loewenstein's character and was simply not credible, whereas the theory of accident could be supported by strong circumstantial evidence.

This evidence began with the destruction of motive. Loewenstein's affairs proved to be in order. He had suffered heavy losses, but he was still a multi-millionaire. His companies were sound. His health was good. His family life – he had a wife and an 18-year-old son – was happy. He was a devout Roman Catholic. He had left Croydon full of plans for the future. He had given no hint of failing courage or will.

Those close to Loewenstein knew that he had been suffering from a slight but tiresome chest ailment. It had not impaired his general health, but it would explain his breathlessness at 4,000 feet, and his desire to open the window to get some fresh air. He had taken off his jacket, and his collar and tie. Then he had gone back to the lavatory.

The mystery of what happened in that aft compartment, out of view of the other occupants of the plane, can never be satisfactorily solved, but an attempt at reconstruction is possible. One theory that has been advanced is that Loewenstein was again overcome by breathlessness and that to get some air he opened the rear door, then collapsed and fell out. This, perhaps, is a little too neat and plausible, and it scarcely rings true. There is, however, a more involved and circumstantial theory which somehow fits the character of Loewenstein.

A door in an aeroplane that serves two purposes is confusing

– a potential hazard. Let us suppose that Loewenstein, engrossed as he nearly always was in business problems, angry with himself no doubt at his unaccustomed physical weakness, made to leave the toilet compartment and return to his seat. He grasped the handle of the door opposite him, leading, as he imagined in that pre-occupied moment, to the cabin. Just possibly it had not been properly shut. As he turned the handle it seemed that the door was unusually resistant to his pressure. With an impatient, violent, impetuous gesture entirely typical of the man, he thrust the weight of his body against it. It opened about two feet, quite sufficient, as he lost his balance, for the slipstream to drag him out. His last despairing shout of warning would be drowned in the noise of the engines.

Did Loewenstein fall – or was he pushed by the relentless pressures of high finance?

The answer must be that he fell.

10

GLENN MILLER

OUTSIDE the block of flats in Waterloo Road, Bedford, stood a jeep with American markings. Presently from the flat there emerged two duffle-coated American officers. One of them carried a large valise. He threw the valise in the back of the jeep, then stepped briskly into the front. His companion had meanwhile jumped into the driver's seat, and with a slam of gears the jeep moved off. Soon it turned past the statue of John Bunyan at the top of the high street and headed north-west along the Rushden Road.

Both men stared apprehensively at the gathering December fog. But it seemed that their journey was to be a short one. Within a mile of leaving Bedford they turned abruptly off the main road and into a winding country lane. In another mile they had reached the flat plateau of Twin Woods Farm, the farmhouse and its red-brick outbuildings taking shape out of the fog as they drew near. Fifty yards from the farmhouse was another building: square, squat and austere. It was an airfield control-tower. A small single-engined plane was pulled up just in front of the tower, and the jeep drove towards it. In the driver's seat of the jeep was Lieutenant Don Haynes, administrative manager of the American Band of the Allied Expeditionary Forces. Sitting next to him was Major Glenn Miller.

The green, level grazing land of Twin Woods Farm had not escaped the covetous eyes of the Air Ministry surveyors when, in 1939, they came to look for airfield sites near Bedford. Farmer John Quenby had been forced to give up some 330 acres of his farm. Behind the farmhouse lay the woods which gave the farm its name. Ahead lay the broad, level acres. The Air Ministry did not trouble to uproot Quenby and his family. They simply built a control tower next to the farmhouse and laid out their runways, dispersing the domestic sites at the farm's edge. Of all the many English farms that were obliterated during the war years, few farmhouses can have maintained their existence, as this one did, in the very shadow of the control tower.

Farmer Quenby and his wife and children watched the airfield grow, saw the training planes taking off and landing, got to know the permanent staff and instructors from the pupils, took a satisfaction in the harvest that had replaced their own. It was an invasion they regretted, but an experience they wouldn't have missed. It left them, when the war ended, with many memories. How could it be otherwise when, from the warm sitting-room – even as late as the nineteen-sixties – one stared through the windows straight into a bleak, derelict monument not fifty yards from one's home – the same old control tower?

It was in front of that tower, one glorious Sunday afternoon in the summer of 1944, that the entire Glenn Miller band had played through its repertoire for the RAF boys of Twin Woods – and incidentally for Quenby and his family. It was in front of that control tower, five months later, that the great bandleader had stepped from his jeep for the last time and lugged his valise into the small Norseman plane that was waiting for him.

To understand why Glenn Miller was on that plane, it is necessary to go back to the days when he was a struggling musician. Born in Clarinda, Iowa, in 1905 of a typical American family, he milked cows for £2 a week as a schoolboy to pay for his first trombone. Through college and university in Colorado he played in various bands, specializing in arranging, and many of the records he helped to make in those early years are collectors' pieces today. His first big break came with Ray Noble, whose American band he helped to organize in the early thirties. Through these years there emerged the image of the tall, studious musician with the rimless glasses whose life was dedicated to the pursuit of a new and distinctive sound – a sound that in spite of his brilliant arranging he could never quite achieve. He became impatient with his role of arranger. 'I'm tired of having things come out different from the way I write them,' he said. 'I want to hear my own ideas and I figure the only way I can do it is with my own band.' But even when he formed his own orchestra in 1937 he couldn't hit the right combination. Bitterly disappointed, he broke the band up. But when he re-formed it the following year success came suddenly. The illness of a trumpeter drove him into a hurried re-arrangement, and the result was an enforced dependence on the clarinet. This, he found, lent to his saxophones a unique and instantly identifiable quality – the sweet, mellifluous sound for which he had searched for so long. It swept the country. Two years later, one out of every three nickels put into American juke boxes were for the records of Glenn Miller. 'Moonlight Serenade', 'In the Mood', 'String of Pearls', 'Tuxedo Junction', one by one his records beat all comers, until in September 1942 his 'Chattanooga Choo-Choo' topped a million copies, the first record to do so for 15 years.

But, as always, Miller was a man who knew just where he was going. When war came, although over-age and exempt from military service through poor eyesight, he enlisted with the idea of bringing his music to the men in the front line. Frustrated at first, his chance came when General Eisenhower, forming plans for the entertainment of the men who were to invade Europe, arranged for a special programme to be beamed to the Allied Expeditionary Forces by the BBC. There were to be three big bands – British, Canadian and American. The bandleaders chosen were George Melachrino, Robert Farnon – and Glenn Miller.

The American Band of the AEF, as it was known, arrived in London under Glenn Miller shortly after D-Day and was accommodated in Lower Sloane Street. But one night in London was enough. The flying bombs had started, and the entire band spent the night in a basement – hugging their instruments. Not a man was risking those. Accommodation was found for them at once in Bedford, about 50 miles north of London, and next night a bomb completely destroyed their billet and shelter of the previous night. The BBC had got them away just in time.

They made their first broadcast on the AEF programme from the Corn Exchange in Bedford on 9th July 1944. Bruce Trent and Dorothy Carless were the vocalists. From then on they worked almost continually on broadcasts and shows.

Miller had a band which would have been quite impossible to keep together in peacetime. There were over 60 performers and the instruments included 21 violins, many of them from the most famous symphony orchestras in America – Boston, New York, Chicago, Philadelphia. Controlling such an agglomeration of talent had its problems, but Miller was a strict disciplinarian and a hard taskmaster. Yet in off-duty hours no one was more approachable. It was a familiar sight in Bedford to see his tall, broad figure five or six places down in the fish and chip queue, and it was a supper he ate in the traditional style – salt, vinegar, newspaper and all.

Life for Miller settled into a pattern. There were shows in London at the Queensberry Club (the Casino), shows at army and air force bases, and continual rehearsing and broadcasting from Bedford, where he shared studios with the BBC Symphony Orchestra under Sir Adrian Boult. Cecil Madden, the BBC producer, got to know Miller well in this period and found him a delightful and magnetic personality. Others thought him sombre and fatalistic, a gloomy go-getter. Cecil Madden noticed this streak of melancholy. One day they were talking at Miller's flat when Miller showed him a scale model of a house in California which he planned to build after the war. It was a film star's mansion, complete with swimming-pools. 'It's a fine home,' said Miller, 'and I can certainly visualize it, but somehow I've got a hunch that I'll never see it. I'll never get back to California.' Madden could not understand this premonition. What could go wrong, even in wartime, for a bandleader like Glenn Miller?

For Madden, working with Miller was a revelation. Miller would move about in front of the band, one ear cocked, one hand beating rhythmically but carelessly, hardly seeming to conduct. Sometimes he would walk away and leave the band playing. Yet always they seemed to be dominated by his magnetism. 'If I hear a musician sticking out,' Miller would say, 'I tell him. And if I don't hear him, I tell him too.' The band knew they were the world's best, and that helped them to put up with Miller's arduous schedules.

When the Allied armies began to break out from their narrow foothold on the Continent, Miller began to dream of taking his band to France. Once the idea had formed in his mind he pursued it with all the single-minded dedication with which he had once sought that distinctive sound. Fresh troops were arriving in Britain from America all the time and Miller was continually playing to them as they got ready to cross into the bridgehead. He began to make them a promise that was to prove a fateful one. 'If you chaps can get there,' he said, 'I'll play you a special Christmas Show from Paris.'

Paris fell all right, but for Miller, getting there wasn't to prove so easy. He went to see Maurice Gorham, seconded from the BBC to head the AEF programme. 'I want to get the band to Paris for a Christmas Show,' he said. 'How about it?' Gorham was sympathetic but pointed out that it couldn't be done. 'Your first responsibility is the radio schedules,' he reminded Miller. 'You can't do those from Paris because there are no studios and no lines, not even through SHAEF. I should forget about it. In Paris you can only play for a few thousand; but your radio audience runs into millions. That's what Ike brought you here for and that's what you've got to do.'

It was all true enough and Miller didn't argue. Yet inside him grew the feeling, unreasoned and unreasonable, that people were trying to stop him from going to France. Somehow there must be a way. A few weeks later he went back to Maurice Gorham. 'Look,' he said, 'suppose we recorded a whole batch of programmes, say for six weeks ahead. If we got those in the can, could we go?'

'As far as I'm concerned,' said Gorham, 'that would do the trick. I'd have no reason to stop you then. But six weeks' programmes in advance! It'd take some doing, wouldn't it?'

And so it did. Week by week all the Miller rehearsals and programmes – from small groups to the full band – were doubled to make the reservoir of recordings that would be needed if the band was to disappear to France for six weeks and still maintain its radio schedules. One programme only was not recorded. This was the special Christmas Show, which was to be broadcast live from Paris on Christmas Day.

At last all the recordings were done, 129 of them, aircraft were laid on to transport the band to Paris and accommodation was booked. On Friday the 13th of December, Miller went to London to clear final details and to say good-bye to Maurice Gorham.

'Everything's in the can,' said Miller. 'We're all set.'

'Good,' said Gorham. 'But just one more thing. See that you get there and back in one piece. And look after yourselves while you're there. We don't want to lose you, you know.'

'*You* should worry!' said Miller cynically. 'You've got the recordings!'

They were due to leave next day, but bad weather delayed them, and Miller began to worry. Suppose after all they were unable to get to Paris in time? The plan had been that Don Haynes should go ahead of them to arrange accommodation and book studios and theatres, but he too had been held up by the weather. There would be a lot of organizing to do, and from what Miller had heard from other artists who had visited Paris, conditions there were appalling. The stars were put in the best hotels, but this had little meaning when there was no heating, no hot water, and very little lighting. Impatient at the delay, Miller made up his mind to get to Paris himself as soon as possible, leaving Don Haynes behind to look after the band and bring them over when the weather improved. It ought, thought Miller, to be possible to get a seat on an operational plane even though large transport planes were still grounded.

It was then, apparently quite by chance, that he met a friend whom he thought might be able to help. His name was Colonel Norman Basselle. He explained to Basselle why he wanted to get to Paris so urgently. 'Sure,' said Basselle, 'if it's as urgent as that we can probably fly you over tomorrow. In fact, I'll come with you.' The orders that had been drafted for Haynes's trip were then cancelled and fresh orders made out for Miller.

Basselle was waiting in the watch-tower at Twin Woods airfield on that foggy Sunday afternoon when Miller and Haynes arrived in the jeep. With him he had an experienced American transport pilot, Flight Officer John R. S. Morgan of the 35th Depot Repair Squadron from Abbots Ripton. 'We're all set,' said Basselle. 'The weather's not too good, but we'll be OK if we get away now.' The RAF had, in fact, advised them not to go.

Miller was in no mood to be held up again by the weather, and he walked with Basselle to the plane. He turned to say good-bye to Don Haynes. 'Get the boys across as soon as you can,' he said. 'I'll have everything set up for them.'

The Norseman was a small utility transport plane with a high wing and a radial engine. It was officially designated a UC-64A. As Miller climbed in he took in the cockpit and cabin layout.

'Where are the parachutes?'

'Parachutes?' Basselle roared with laughter. 'Do you want to live for ever?'

The cabin door closed and Haynes walked back to the jeep. The fog was drifting in more thickly now, and as the Norseman taxied out towards the runway its outline became indistinct. Soon it disappeared altogether in the mist and fog.

There had been no flying by RAF planes at Twin Woods all day; conditions weren't good enough for training flights. But the field wasn't technically closed. If the Americans liked to take the responsibility, that was up to them.

As far as anyone knew that evening, the great Glenn Miller was safe and well, right at the top of his form. His band played as usual that night on the AEF programme at 8.30. It sounded better than ever. The announcements were made in the same clipped American drawl by a voice that was quite unmistakably Glenn Miller's. It was, of course, a recording.

It was not until two days later that the band got away in their four-engined transport aircraft. When they got to the airport at Orly, just outside Paris, their first questions were for

Glenn Miller. They were surprised that he wasn't there to meet them. Perhaps he had left a message for them. No? Then how could they contact him? Nobody seemed to know.

'When did he get here?' asked Don Haynes.

This question, too, was met by a blank look and a shrug of the shoulders. No one knew the answer. Indeed, no one remembered seeing or hearing anything of him. Perhaps he had landed somewhere else.

It was at about this point that Haynes began to impart to his questions an urgent, insistent ring. It was some time before anyone could be convinced that Miller was missing, but slowly the truth was uncovered. Miller and Basselle, together with their pilot, had taken off that Sunday afternoon into the Bedfordshire fog and disappeared.

A search was begun along the route from Bedford to Paris, a search which revealed nothing. The most likely explanation seemed to be that Miller and Basselle had been forced down somewhere in their Norseman and that they would come striding in large as life at any moment. On this assumption, arranger Jerry Gray took over the baton in Paris and the band started rehearsing.

The atmosphere in Paris was eerie and unreal as rehearsals went on. In the evenings at their hotel, without heating and by the light of candles, the band, apprehensive and fidgety, discussed their situation. The recordings they had made in London and Bedford were still being pumped out by the BBC. Glenn Miller's voice, normal and matter-of-fact as ever, came to them from their portable radios in odd reproach. Most of the time they switched off.

In London the BBC quickly prepared duplicate recordings in which the band numbers were retained but the announcements were taken out, against the moment when the Americans should decide to release the news that Glenn Miller was missing. But the Americans had suspended all casualty lists until after Christmas, and meanwhile the BBC had no alternative but to keep up the deception. Thus for a further week the voice of Glenn Miller, now almost certainly stilled for ever, went on bringing its warm encouragement to the men of the

AEF, that encouragement which, to quote Eisenhower, was the greatest morale builder in the European theatre next to a letter from home. All the great tunes associated with Miller – 'Moonlight Serenade', 'Pennsylvania 65000', 'American Patrol' – were broadcast in that last agonizing week. Then at last, as wild rumours that Miller had been captured or had defected to the enemy began to filter through the theatre, and in the face of the promised broadcast on Christmas Day for which everyone was waiting and for which no recordings had been made, the Americans were forced on Christmas Eve to release the news that Miller was missing.

The announcement started another flood of rumours which has left its residue to this day. No trace of the plane or any of its occupants was ever found, and the conclusion must surely be that it came down in the fog somewhere in the North Sea. But the rumours persisted. It seems that nothing is more certain to breed legend and far-fetched speculation than sudden disappearance without trace. Many people indeed found it hard to believe that, in an area where the Allies had complete control of the skies, a serviceable plane of the known reliability of the Norseman could vanish so utterly, with nothing whatever known of its fate. Surely somebody must know something. Could the disappearance have been deliberate? Miller, it was suggested, might have been acting as a secret agent; his disappearance might have been engineered. One fanciful rumour had it that he was a black marketeer who had made a wartime fortune and then chosen to disappear, persuading his pilot to fly to Germany, or to some spot behind the Russian lines. It was even said that he was a psychopathic case and that the authorities knew perfectly well where he was. Another report said that he was alive but that his face and body had been terribly mutilated in an air crash. But one by one all these rumours were shown to be demonstrably false and ridiculous. Eventually Miller's death on active service was presumed.

One thing that would surely have silenced the rumours would have been the publication of the findings of the court of enquiry into the circumstances of Miller's last flight – if such an enquiry was held. The enquiry would have been expected to establish whether the Norseman was on a properly authorized flight, and whether it reported its take-off time, destination, and estimated time of arrival to the appropriate airways control. If the plane then failed to arrive, overdue action could and would have been taken. But no one seems to have known about the disappearance until Don Haynes began his enquiries at Orly. Or was this an erroneous impression gained by Haynes and perpetuated in subsequent, ill-informed accounts of the affair? Had enquiries in fact already started, had air/sea rescue searches been made? If not, was it through some oversight in airways control, or did John Morgan, an experienced and reliable transport pilot, fail to pass the routine departure signal? Could it perhaps be that the flight was actually made against orders, under persuasion from Miller?

All these questions have remained unanswered, though the answers to them must surely be known. If there was no full enquiry, what was the reason? Was there after all something to hide? Apart from a brief statement which included the aircraft type, the pilot's name, unit and base, and the date on which pilot and plane were reported missing, the Air Force Historical Office in Washington was unable as recently as 1963 to answer questions, and they further stated that no more information was available on their files. In view of characteristic American generosity in providing information, this statement must surely be accepted without question. It deepens the mystery still further.

There is, then, the suspicion, the probability even, that the flight was not properly authorized. And there is another rumour, particularly distressing to British ears, which also persists. It is still repeated, in a confidential whisper, by people who say they knew somebody who had it from a chap in the Eighth Air Force. . . . It is that Glenn Miller was shot down by the British. The rumour, naturally enough, has gained special credence in America. Miller's plane, one is told, was shot down by an Allied fighter, an Allied fighter which for once wasn't American. In plain words, by a British plane.

Fortunately it appears that this distressing rumour is unfounded. Not a single contact was reported on 15th December by any of the British commands and units flying in the area. Bomber Command, Fighter Command, Coastal Command, 2nd Tactical Air Force, all reported nil contacts and nil claims. Twenty enemy aircraft were sighted in the battle area during the day. Nineteen of them were jets. The twentieth, according to the sighting report, was a Ju 88. It was attacked and damaged, but it broke off the engagement and disappeared. Its attacker was an American plane of the famous Black Widows squadron. It would be ludicrous to suggest that this squadron could confuse a Norseman with a Ju 88.

So, unless the Americans have some secret intelligence of which the British are ignorant, we can acquit the Allied Air Forces of making the mistake of shooting down Glenn Miller. The probability is that the bandleader, doing his duty as he saw it and determined to keep his promise to make the Christmas broadcast from Paris in spite of the personal risk involved, took a chance on the weather. In doing so he probably defied authority. The weather was bad when he took off, and it rapidly got worse. No signals were reported to have come from the plane after take-off, so it may be that the small voice-radio that the plane would carry failed or was inadequate. Unable to get any check on his navigation in the fog and without radio communication, the pilot must have found himself running short of fuel somewhere over the North Sea. Eventually the plane must have been forced down. No one could survive for long at that time of year in those icy waters.

The band duly made the promised broadcast from Paris on Christmas Day – but without its devoted leader. As the Glenn Miller band it continued to play for a good many months, but its genius and driving force were gone. That casual professorial personality whose influence on the band had sometimes seemed negligible was in fact irreplaceable. Without it, the band was never the same.

11

'WRONG-WAY' CORRIGAN

'SEE here, Corrigan – you can't take off until morning.'

'But I'm all set to go at midnight, Mr Behr.'

'Sorry, that's the ruling. In any case I'm not having you risk a night take-off with that load of gas.'

'It's my neck, Mr Behr.'

'Sure, sure. But you'll still have to wait until clearance comes through.'

Douglas Corrigan, the young man who waited through the night at Floyd Bennett Field, New York, for permission to take off on a solo non-stop cross-country flight to Los Angeles, was a very boyish 31. His 165-hp single-engine Curtiss Robin monoplane, ancient, rickety, and half a ton overloaded through additional fuel tanks, had an endurance of about 30 hours at a cruising speed of a hundred miles an hour. On paper he would have a fair margin for the return flight to California, but in practice he had had only four gallons to spare – about 20 minutes' flying – at the end of his outward leg to New York.

Corrigan's idol was Lindbergh – had been ever since he had worked eleven years earlier, in 1927, as a mechanic for the Ryan company, preparing and servicing the plane that Lindbergh flew on his great solo crossing of the Altantic. Corrigan's ambition had always been to emulate that flight. He would certainly have been attempting the Atlantic crossing now, but permission had been consistently refused. Long-distance ocean flights in private planes had not been popular with the authorities since the disappearance of Amelia Earhart in the previous year.

Even the non-stop flights from California to New York and back had required tact and careful planning, not to say subterfuge. Failure on the outward flight would have meant cancellation of the licence for the round trip, so Corrigan had been careful not to declare his intention of starting out, and to take petrol on at two different Californian airfields so as not to arouse the suspicion that he was embarking on a very long flight. He had taken off unheralded, and if he had failed he would simply have kept quiet and tried again. Now, with the first leg successfully behind him, he did not need to be secretive about the return trip.

The news of his non-stop flight from Los Angeles had been ferreted out by a newspaper man and given a little mild publicity; he had got his name in the paper, and made a broadcast on the NBC network. But soon his existence was forgotten. One person who heard about the flight of the young mechanic, however, did not forget him. Ruth Nichols, the world-famous woman aviator, remembered that he had once worked on her plane, and she offered to lend him a parachute for the return trip. He thanked her but said he had no room for it. 'In any case,' he said, 'the plane is all I have and if it falls to pieces I may as well go with it.'

On Saturday, 16th July 1938, exactly a week after his arrival in New York, Corrigan had his plane ready for the return flight, and the weather forecast was good. He announced his intention of taking off at midnight, and it was then that obstacles were put in his way by the airport manager.

'Have you got a special licence to fly this plane?'

'Yes, I have. You can check that with the Bureau.'

'I'll do that.'

'How about the take-off then, Mr Behr?'

'You heard the ruling – you can't go till morning. I'm not even sure we'll let you go then.'

But at four o'clock that morning the airport manager had relented. No doubt he had checked meanwhile on the licence. 'It's OK for you to get ready,' he said. 'You can take off at dawn.'

Corrigan and an airport mechanic wheeled the plane out of the hangar and filled the tanks. Corrigan cranked the propeller, and soon the engine was ticking over. He examined the motor with a flashlight to make sure everything was in order. He made a similar check of the cockpit, then walked across to the airport manager's office.

'Thanks, Mr Behr – I'm all set to go now. Which runway do I use?'

'Any one you like. But don't take off towards the built-up area at the west of the field.'

Corrigan examined the windsock. There was very little wind. 'I'll take off on the long runway,' he said, 'west to east.' He would need a good run with his heavy load.

First light was only just seeping in under the horizon as Corrigan aimed his plane down the runway. There was considerable ground haze, with patches of fog, but he could just make out the edge of the concrete runway and keep the plane on the right heading. It took a thousand yards to come off. Even at 50 feet the controls were sloppy and sluggish, and he realized he would need more speed and height before he could safely turn on course. He flew on over Long Island in an easterly direction, straight and level, while the plane responded reluctantly and he slowly gained height.

By the time he reached 500 feet, the ground below was obliterated by fog. He began a gentle turn to starboard, intending to come right round on to a westerly course, and his eye took in the compass for a moment to check the amount of turn. He was dismayed to find that it wasn't working properly. The liquid inside the compass had somehow leaked away.

He had checked the entire plane over the night before, but this was something he must have missed in the darkness. Fortunately there was a second compass, down on the floor near his feet, which he had set before take-off for a westerly course. All he had to do was to continue his turn until the parallel lines on the second compass lined up correctly. When this point was reached he settled down on course, still climbing steadily, all ground detail still hidden by fog.

In fact, Corrigan had perpetrated one of the oldest known navigational howlers; he had misread the second compass, and he was flying on reciprocal, east instead of west, straight out to sea.

The fog that clung to Long Island did not altogether blanket the view from the ground. Corrigan himself could recognize nothing, but several people caught occasional glimpses of his plane. Among these were a handful of friends and one or two officials at Floyd Bennett Field. The suspicions of the officials were quickly aroused. The plane was climbing strongly now, heading quite deliberately out towards the Atlantic. The history of Corrigan's rejected applications for an Atlantic flight was recalled. Under the pretext of a return flight to Los Angeles, it seemed that he was sneaking off on the forbidden crossing.

It occurred to officials that it might be more than coincidence that the flight from New York to Ireland, with the prevailing winds, would burn little more fuel than one from New York to Los Angeles. But Corrigan's friends discounted this theory altogether. It was true that in the previous year, irked by the persistent refusals of the Bureau of Air Commerce, Corrigan had threatened to do exactly this. 'Why not land at Floyd Bennett Field late one evening,' he had said, 'when senior officials have gone home, fill up with gasoline, and just go?' No one could hang him for it. That way he would get no weather reports, but he would have to take a chance on that. With this idea in mind he had set out illegally from California for New York, his plane unlicensed, landing on derelict war airfields or on open ground, but by the time he had reached New York it had been too late in the year to attempt an Atlantic crossing in a light plane. He had had no de-icing gear, and he would have been bound to meet severe icing conditions for much of the way. He had had to go back to Los Angeles, where his plane had been grounded. That had been the end of Corrigan's plan for flying the Atlantic – so far as anyone knew.

Corrigan had given his friends not the smallest hint this time that he might be about to attempt the Atlantic crossing. He had no radio. He had no weather data for the Atlantic. He didn't even have any maps, other than a Great Circle map of the United States, marked out with his planned trans-Continental route via Memphis and El Paso. Apart from a few fig bars and some chocolate he had no food. He had no warm clothing, no water, no clearance papers and no passport.

Then there was the question of the reliability of his plane. Corrigan was an experienced mechanic and he would be well aware that his machine, safe enough perhaps for flying over land, where a successful forced landing was almost always feasible, was quite unsuitable for a long sea crossing. Corrigan kept his plane in good shape, but he had arrived in New York with a leaky petrol tank, and rather than remove the tank and weld it, which would have held him up for another week, he had decided to risk it. It was inconceivable that he would have taken such a chance if he had intended to fly east.

Both friends and officials recalled Corrigan's care over the weather. He had studied the cross-country route conditions at the nearby weather centre at Mitchell Field, noting especially the strong and variable winds – winds that might be against him in the early stages of an Atlantic crossing, giving him no margin at all. He had been given no weather information on the Atlantic. If the easterly flight had been his intention he had carried out a gigantic and probably suicidal bluff.

By his action at the beginning of the flight from Los Angeles – taking on fuel at two different airfields, and keeping his own counsel about his intentions – Corrigan had shown himself capable of subtlety and discretion. But his friends had been let into the secret. It was preposterous now, for this much bigger enterprise, that they could have been left in the dark. Indeed Corrigan was of such a frank, open disposition that his friends did not believe him capable of such a deceit. The New York authorities, however, took a different view. Soon after Corrigan's take-off, after the plane had disappeared into the eastern fog-bank and no news had come of any sighting inland, they warned all shipping in the Atlantic to look out for Corrigan's machine, and took it that they had been duped.

Meanwhile Corrigan was two hours out of New York, flying above the cloud now, having climbed to 3,000 feet. He was flying a Great Circle course which would bring him over numerous landmarks if he kept on track. He estimated his ground speed as just over a hundred miles an hour. Soon through a gap in the clouds he caught glimpses of a city which he took to be Baltimore. In fact, as he would have realized if the clouds had not obscured the waterfront, he was over Boston, Massachusetts, heading out to sea.

He was flying now between two solid layers of cloud. There was nothing at all to be seen, either above or below, and there was nothing he could do but sit tight and follow his compass course. Hour by hour, as the fuel load lessened, he was steadily gaining height. Eight hours after take-off he was at 4,000 feet. The cloud-layer below him was rising and he was scudding over the tops of it, giving a comforting sense of speed and progress. It was an odd sort of progress. Instead of being, as he

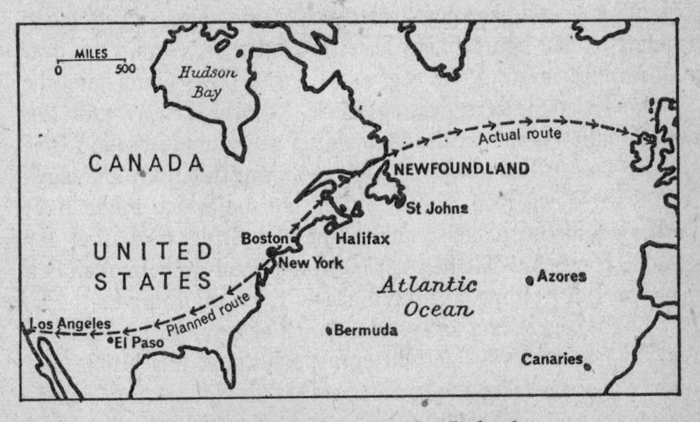

6. New York–Los Angeles–Ireland

imagined, above the plains of Kentucky, north of the Appalachian mountains, he was over New Brunswick, 800 miles out from New York.

It was not until he had been airborne for ten hours that he caught another brief glimpse through the cloud-bank. It did nothing to disabuse or dismay him – in that moment he was crossing the northern tip of Newfoundland and there was no sign of sea. He was unable to get a pinpoint, and he could not recognize the terrain, but his attention was now distracted by something else. He noticed that his feet seemed to be extremely cold, which was not a thing he normally suffered from, and looking down he saw that fuel from the leaky gasoline tank was seeping into the cockpit, soaking his shoes. In the chilly atmosphere, the dampness and evaporation had had a refrigerating effect, making his feet feel like ice.

The seepage increased his anxieties about endurance, which he had hitherto considered adequate. The actual loss of fuel as yet must be small, but he noticed with some uneasiness that the leak appeared to be getting worse. However, he was confident of being able to make a forced landing in emergency, and he was more disturbed by the danger of fire.

The cloud was building up beneath him and he was still gaining height, keeping just clear of the lower cloud layer. After twelve hours flying he estimated that he must be over Memphis, starting the leg to El Paso. Two hours later he judged that he was passing Little Rock, Arkansas, still less than half-way across the States after 14 hours in the air. The cloud was still an unending layer-cake, pressing up from below, pressing down from above, hiding the sun which might have aroused his suspicions, hiding the shattering truth that lay 6,000 feet below. He had left Newfoundland far behind and on his present heading still faced nearly 1,700 miles of sea.

Night was falling and the sandwich of clear air between the cloud-layers slowly blacked out. No horizon was visible and Corrigan was forced to concentrate on his rudimentary instruments – turn and bank indicator and air speed was all he had – throughout the night. Hour after hour he flew on, imagining that he was crossing the vast plateaux of Texas and approaching the Guadalupe mountains before passing El Paso. Sometimes there were fleeting breaks in the cloud and he peered below for lights, for the white ribbon of roads or the shadowy curves of a river, but he was not greatly surprised when he saw nothing.

His chief concern was the leaking fuel tank. He could hear the gasoline slopping about in the bottom of the plane, and when he shone his torch at the cockpit floor he saw that the liquid was an inch deep. He had no idea what fuel he had left. There were no gauges for the extra tanks. For all he knew he might have lost twenty or even fifty gallons, might even face a forced landing, down there in the darkness, perhaps among the mountains, at any moment. Worse still was the numbing, insistent fear that the plane might catch fire. It was of all-wood construction, and if the fuel leaked out through the cockpit floor near the exhaust pipe and ignited, the fuselage would be a furnace within seconds. He almost wished he had accepted that offer of a parachute.

He had to do something to get rid of this carpet of gasoline that covered his feet. The only possible answer seemed to be to punch a hole in the cockpit floor on the opposite side to the exhaust pipe and hope the fuel would drain away safely. He took a screw-driver from his tool-kit and stabbed a hole in the wooden floor, on the right-hand side of the plane. Soon the cockpit floor was almost dry.

It was pointless to keep the engine at the normal economical cruising setting of 1,600 revolutions per minute. The best thing surely was to make use of the fuel before it leaked away. He increased the engine revs to 1,900, burning up his fuel at a much faster rate but making a better air speed. It had become a race to get as far as he could, to his destination if possible, before his fuel gave out.

The darkness was impenetrable, a black wall that he drove through interminably. Yet somehow the night seemed mercifully short. Dawn came slowly, the volume of light increasing gently, with still no sign of the sun. But the daylight was a blessed relief, banishing the terrors of the night. The cockpit was clear of petrol, and he felt much more confident about a forced landing now.

During the night he had climbed to 8,000 feet, but he was still flying only just above the lower layer of cloud, almost scraping it with his fixed undercarriage. Ahead of him the cloud-free corridor was closing as great masses of cumulo-nimbus piled up to a height of 15,000 feet. Climbing above those mountainous cloud-peaks was beyond the capacity of his plane, besides being wasteful of fuel, and he had no alternative but to enter the storm-clouds and fly on instruments. This he did for the next two hours, in visibility that was reduced to nothing by driving rain which coated the windscreen, glistened on the wings and seeped through the cockpit hood. All the time the plane bucked and reared, jerking the control column out of his hands and making instrument flying an exhausting and frightening task.

After a time the rain turned to sleet. He could feel the drop in temperature. He seemed to have passed through the worst of the storm but now he was beset by a new danger. The wings were icing up. His de-icing gear was of the crudest kind imaginable – an 8-foot stick which he poked out of the cabin window and prodded at the rapidly-forming coating of ice. He managed to knock some of it off, but he was obsessed by the worry that his instruments would ice up, leaving him with no check on his air speed. He had no more than a rough – and completely illusory – idea of where he was, and he feared that he might be among the mountains to the north of El Paso. But he feared the icing even more, and he put the nose down and began a steep descent through the cloud.

He was down to 3,500 feet when he emerged from the cloud. The mountains that he had peered for so nervously all the way down were nowhere to be seen. He was puzzled and nonplussed to find that he was over the sea. He must have flown right across the American continent without realizing it. He must be over the Pacific. How far out to sea could he be? He looked back over his shoulder, then swung his gaze systematically through 360 degrees. There was no land in sight.

His first reaction was to turn back at once for the Pacific coast. With the increased engine revs he must have made a much faster ground speed than he had imagined. The winds, too, must have been of less velocity than forecast. Yet even taking these two factors into account he could not get over his surprise at crossing America in 26 hours.

Whatever the situation was, he had to make a decision at once. He still had no check on his fuel state, and if he didn't find land quickly he was facing disaster. As he prepared to turn back for the Pacific coast, which he knew must be somewhere behind him, he glanced down at the compass. He was still using the check compass, the one at his feet. It was slightly in the shadow, not too easy to read, but now the light was better he could see it more clearly. He was almost afraid to look at it. The suspicion that he might be flying a reciprocal course, that he might be somewhere over the vast wastes of the Atlantic, was hardening into certainty.

A closer scrutiny of the compass confirmed his worst fears, and he almost blacked out as the blood rushed to his head with the knowledge of the fantastic beginners' error he had perpetrated. For more than 26 hours he had been following the wrong end of the magnetic needle, flying east instead of west.

His mind spun crazily, unable at first to regain any sort of equilibrium or orientation, lost in the contemplation of a million square miles of ocean. He had to fly the plane, he had to scan the horizon for the first hoped-for glimpse of land, so as to turn towards it at once, and he had to try to plot his course backwards across the Atlantic, working out where he might be in his head.

His mind remained in a turmoil of confusion, boggling at the complexity of the navigational problem that faced him. In the end he surrendered to the pressures of inertia and decided that he simply didn't have the ability or strength of mind to interfere with the course he had flown for so long. With luck he hoped that his present heading might bring him in somewhere over Ireland. How soon that might be, and whether his fuel would last out, he had no idea, and he tried not to think about it. But in fact, the prevailing westerly winds had been helping him for many hours, and he was not far short of the Irish coast.

Within a few minutes he caught sight of a small trawler almost straight ahead and he put the nose of his plane down and dived gently towards it. Flying low over the water, he passed close to the trawler but he could see no signs of life on board. Everyone must be asleep in the cabins below. This was disappointing, but he could not afford to circle until they woke up, so he flew on straight ahead. Such a small trawler was not likely to be ocean-going, so he must surely be near land. Unless he had made a second outrageous error, that land must lie somewhere ahead.

Perhaps the seamen on the trawler had been having lunch. The thought activated his stomach juices for the first time. He realized that he had eaten nothing since Saturday evening, and it was now Monday afternoon. All he had on board were the fig bars and some chocolate. He was chewing these ravenously when the amorphous switchback of clouds on the horizon suddenly seemed to resolve itself into a sharper outline. There was a tinge of colour about them, too, a greenness that could only mean land. He was seeing for the first time the green hills of Ireland. After 27 hours' flying in the wrong direction he had flown the Atlantic.

Forty-five minutes later, without sighting anything more than a few scattered farmhouses and villages, he came to another coastline and realized that he flown across Ireland. The following wind and the light fuel load had sent up his speed considerably. On this eastern coastline he saw a large city, but he could see no sign of an airfield, so he turned south and began flying down the coast. Soon a small military plane, a fighter, came in close to take a look at him. Now, thought Corrigan, he'll lead me to the nearest airport. But the fighter plane dived away and disappeared, and it was not until he reached another large coastal city that he had any confirmation of his position. This time he found the airport, which was clearly marked 'Baldonnel'. He remembered, from his planned crossing of the previous year, that this was the airport for Dublin. His fuel seemed to be lasting out, so he circled carefully to check the wind direction and study the field. Then he came in to a perfect landing.

'My name's Corrigan,' he told the airport official who came out to meet him. 'I've just come from New York.' He expected this to cause a sensation.

'Yes – we know.'

'You know? How?'

'New York warned us about you. They saw you start out and guessed you were making for Ireland. And we had a report that your plane had been seen up north.'

At first everything went smoothly. Airport officials, Customs men, military authorities and government departments all showed great toleration and kindness. The only time when he ran into difficulties was when he tried to explain how the flight had gone wrong. Whenever he got to this point in the story, the party had a habit of breaking up. It was the same when he met the American ambassador, and again when he was introduced to Mr De Valera, the Eire Prime Minister. As soon as he got to the part about misreading the second compass, everyone started to laugh. 'Now tell us the *real* story,' was everyone's reaction.

But Corrigan insisted that his Atlantic flight had been unintentional. 'It sure does show,' he told his countrymen in a broadcast that night, 'what a fool navigator a guy can be.' That's my story, was his attitude, and I'm sticking to it. He showed a refreshing reluctance to capitalize his flight, refusing all easy-money offers, and when a reporter offered him 500 dollars for an exclusive story he told him he could have it for nothing.

The sceptical American nation loved Corrigan for what they made up their minds was a gigantic leg-pull, just the size of joke they liked. He had cocked a snook at authority, too – and they loved him for that. The unsophisticated, immature young man had outsmarted the experts, and in so doing convinced all Americans – and anyone else who cared to doubt it – that their image of themselves was real.

'Most amazing thing I ever heard of!' was the reaction of the American ambassador in Dublin. 'But what are you going to do when the technical experts get to questioning you?' Corrigan certainly had some explaining to do – it was remarkable that he never saw the sun in 26 hours, odd that he didn't notice the considerable difference in the hours of daylight flying east from flying west, strange that the reciprocal course should bring him in exactly over Ireland. Some of this seemed to be more than mere coincidence. But Corrigan stuck to his story, and not even the shrewdest district attorneys or the most elaborate lie detectors could break it down.

Did 'Wrong-Way' Corrigan make an honest mistake, for which a doubting world should have given him the benefit?

Or did he bring off one of the most spectacular hoaxes of all time?

The answer remains as equivocal as it was immedately after the flight, when the United States Flag Association awarded Corrigan its medal for 1938 – and he was unanimously elected an honorary member of the Liars Club of America.

12

THE DUKE OF KENT

THE man at the driver's seat of the polished black saloon turned to look back as the wheels of his car crunched down the gravel drive. In the doorway of his home, sheltering from the hot August sun, a woman and a child were waving. It was a scene no different from that enacted hundreds of times daily in wartime Britain. The precious days together had been savoured. Now the man, home for the christening of his third child, was returning to duty.

The wrist of the uniformed arm that waved in salute carried the inch-thick ribbon of an air commodore. High rank indeed, but not exalted. Yet he was a king's son. The woman who stood in the doorway was Princess Marina, Duchess of Kent. The child was the six-year-old Prince Edward.

The Duke of Kent, fourth son of King George V, had been born in 1902. As a young man he had spent ten years in the Royal Navy. Artistic, Bohemian, unconventional, he was altogether different in outlook and temperament from his brothers. He loved and understood music, was an accomplished pianist, and was something of an antiquary. Allied to this were a quick sense of humour and a charm of manner which captivated everyone.

In the summer of 1939 the Duke was preparing to leave England to take up the post of Governor-General of Australia. But when war came he determined to play a more active part. After a spell on Intelligence duties at the Admiralty he transferred to the Royal Air Force, relinquishing his honorary rank of air vice-marshal at his own request so that he could fill a group captain post in the newly formed Welfare Branch.

Touring RAF stations at home, he would arrive without ceremony, unannounced and sometimes even unrecognized, driving his own car. He watched the flying crews at work, attended their briefings, and awaited their return from operations; and he showed equal interest in the welfare of the ground crews, encouraging them to make their own criticisms and suggestions. In 1941 he flew to Canada to see the Empire Air Training Scheme at work, becoming the only member of the Royal Family to cross the Atlantic both ways in a bomber. And it was in Canada that one of the warmest tributes was paid to him. Fittingly, it was a personal one, committed to the pages of a private diary. The diarist was a senior Canadian officer, Air Member for Personnel in the RCAF, who met him during the tour. His name was Air Vice-Marshal Harold Edwards, or Gus Edwards, as he was universally known.

'I have seen and met,' wrote Edwards, 'many dear people. . . . But in all my 49 years I have not felt the warmth I do tonight, for I have seen and felt something that I have never seen or felt before. A new character distinct and certain has been added to my memories. . . . Perhaps I shall be forgotten before this day is out, but *I* shall not forget.'

On 4th August 1942 the month-old Prince Michael had been christened at St Peter's, Iver. One of the godparents, by proxy, was President Roosevelt, with whom the Duke had stayed at the White House on a brief visit following his Canadian tour, renewing a friendship that went back to 1935. Now, after three weeks' leave, the Duke was off on another overseas welfare tour, this time to Iceland. He drove to Euston, and there he was joined by his private secretary and close friend, Lt John Lowther, RNVR, a grandson of Lord Ullswater; by his acting air equerry, Pilot Officer the Hon. Michael Strutt, son of Lord Belper; and by his batman, Leading Aircraftman Hales. Michael Strutt had been sent for at the last minute because of the illness of the appointed air equerry. The stationmaster at Euston saw the party off on the train for Inverness.

Meanwhile, at half-past three on the previous afternoon, a Sunderland flying-boat of No. 228 Squadron had left its base at Oban on the west coast of Scotland and flown north-east across Inverness-shire to Invergordon, the naval base on the protected inlet of Cromarty Firth. The crew knew that something special was on, but it wasn't until the evening of the 24th, when the Duke of Kent and his party arrived at Invergordon, that they learned who their passengers were to be. Invergordon had been chosen for the rendezvous because it was the most accessible of the flying-boat bases by rail from London.

The Sunderland pilot, Flight Lieutenant Frank Goyen, a 25-year-old Australian from Victoria, of exceptional ability, athletic, steady and dependable, had been specially chosen for the task. Giving up an office job in Australia in 1938 to join the RAF, he had spent the first two years of the war patrolling the Mediterranean and the North and South Atlantic, amassing nearly a thousand operational hours in the process. Once, in June 1940, when shadowing the Italian Fleet, he had alighted on the sea when his petrol ran low, keeping the ships in sight all night without being spotted and then taking off next morning and flying back to his base with a detailed report. He was the Sunderland pilot *par excellence*.

With him Frank Goyen had his regular crew, equally trusted and competent. They numbered ten in all and included a second pilot, two radio operators, three gunners, a navigator, an engineer and a fitter. The second pilot, too, was an Australian, and another crew member was a New Zealander; the others were British. In addition, as a courtesy and to mark the importance of the flight, the squadron commander, Wing Commander T. L. Moseley, also an Australian, was aboard. Moseley was a former Cranwell cadet who had been on flying-boats since 1934. Before taking over 228 Squadron he had been on the staff of the Deputy Director of Training (Navigation) at the Air Ministry.

Tuesday the 25th was a typical wet August day – a day of almost continuous rain. The fine weather had broken and storms and low cloud covered the British Isles. Northern Scotland was no exception, but the weather was not regarded as bad enough to postpone the Duke's flight. Cromarty Firth was clear, the cloud-base being about 800 feet, and the forecast for the Faroes, which the Sunderland would overfly *en route* for Iceland, was that the weather was improving. At half-past twelve Moseley, Goyen and the rest of the Sunderland crew went out by tender to their aircraft.

The flight to Iceland, a distance of nearly 900 miles, would take about 7 hours. The aircraft was a Mark III Sunderland, registration number W4026, built by Short Bros at Rochester. The cruising speed was about 110 knots and the endurance nearly 12 hours. Facing a long sea crossing, with the possibility of being forced to return to a base in Scotland if the weather at the destination deteriorated, the flying-boat had a full load of fuel. There would be 15 people on board, and in addition the Sunderland carried depth charges in case submarines were seen *en route*.

Shortly before one o'clock, the tender carrying the Duke of Kent and his party set out from Invergordon pier. The Duke himself, lean and extremely fit, had a friendly smile for the crew as he came aboard. 'Just like his pictures,' said the Scottish tail-gunner, Andrew Jack. When the tender had got clear on its way back to the pier, the four Bristol Pegasus engines were started one by one. Then the Sunderland slipped the buoy to which it was moored and began to taxi slowly into Cromarty Firth.

In the shelter of the Firth the water was smooth, almost too smooth, since the ripple of waves that normally helped a flying-boat to take off would be missing. It was a long take-off run, but at length Goyen pulled the hull clear of the water, and for the next few minutes they flew at low level in an easterly direction along the Firth, skimming the water, until they passed between the 450-foot high Sutors, the two precipitous rocks that guarded the entrance to the inlet, and emerged into the open sea.

Because of the heavy load it was not intended that the Sunderland should turn at once on a direct course for Iceland. That would mean crossing the fringe of the North-West Highlands, where there was much high ground, with several spot heights above 3,000 feet. The climbing rate of a fully-loaded Sunderland was not quite good enough for that. Goyen had been briefed to follow the coastline in a north-easterly direction for 85 miles, keeping out to sea, before turning to

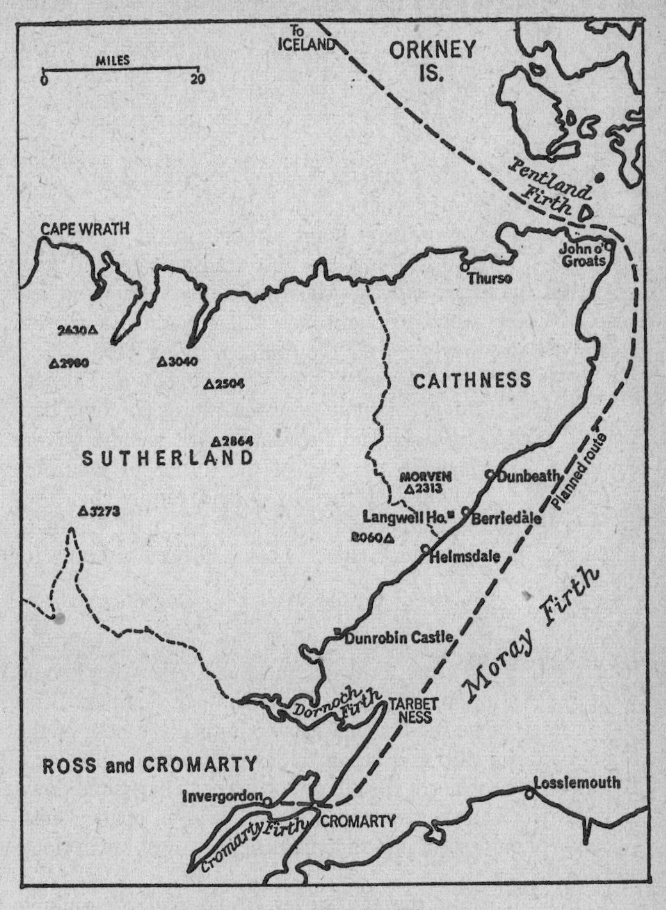

7. Cromarty Firth to Caithness

port off John o' Groats to pass through Pentland Firth, south of the Orkneys. From this point, course would be set for Iceland.

At the mouth of the Cromarty Firth estuary, Goyen turned 45 degrees to port and began the coast-crawl north-eastwards, climbing as he went. The crew took up their positions and called Goyen.

'Tail-gunner calling captain. Testing intercom. Over.'

'Captain calling tail-gunner. Loud and clear.'

And so it went on from point to point in the flying-boat, until all the crew had made a test transmission and been answered by Frank Goyen. Meanwhile the Sunderland had reached about 1,000 feet and was still climbing slowly but steadily. Soon they were off the promontory of Tarbet Ness. To the left the land receded into the great bay of Dornoch Firth before turning obliquely north-eastwards to come back parallel with the Sunderland's track. But the cloud-base was lowering. They caught glimpses of the turrets of Dunrobin Castle, ancestral home of the Sutherland family, but they were still off the Sutherland coast and south of Caithness when they flew into thick cloud. They had been airborne for 25 minutes.

On a coast-crawl of this kind, with high ground not far away, it was always reassuring to see the sea below, with the long slanting coastline away on the flank. Also they would want to get a pinpoint before running through Pentland Firth and setting course for Iceland. Goyen pushed the nose of the Sunderland forward and began to descend through the cloud. But as they went north the cloud got lower. They came down to 1,200 feet and they were still shrouded in an opaque mist. Goyen kept on descending. In a moment they must break cloud and see the sea.

Andrew Jack, in the tail turret, could feel the sinking motion as the Sunderland descended. He guessed what was happening – they were going down so that they would follow the coastline.

In fact, through some unaccountable mischance, the Sunderland, although still over the water, was drifting in towards the coast at an angle and on its present course would cross the coastline a mile or so north of the Sutherland-Caithness border. Inland there was much high ground.

Goyen, evidently satisfied that he was still over the water, went on descending. He actually crossed the coast above the Duke of Portland's estate near Langwell House. As he continued northwards, the coastline swept away to the right, leaving him well inland.

He was now crossing the rugged, sparsely-populated fringe of an enormous area of purple mountains. Beneath him lay a stretch of land that was at once a pasture for sheep and a refuge for stag. Occasionally a white-washed cottage broke the monotony of the hillsides, but Goyen and his crew were still flying in thick cloud and they saw nothing of this. Ahead of them, directly in their path, lay a sharp eminence 900 feet high known locally as Eagle's Rock. At its present height the Sunderland would just clear it.

Even the hillsides below Eagle's Rock were shrouded in mist. But two men were down there on those hillsides, rounding up their sheep. They were farmer David Morrison and his son Hugh. Somewhere in the mists above them they could hear the roar of aircraft engines, but they could not see even a shadow of the plane's outline. The noise had dwindled but was still in their ears as the Sunderland passed over Eagle's Rock.

The fluctuating heights of the ground below were causing considerable turbulence in the Sunderland. An upcurrent cushioned the flying-boat as it passed narrowly clear of the peak of Eagle's Rock; a precipitous down-draught followed. On the far side of the rock summit was a rounded shoulder 800 feet high, towards which the Sunderland was thrown.

The impact was no more than a glancing blow, but the flying-boat shuddered to its rivets. It ricocheted into the air, turned over uncontrollably on its back, and crashed with disintegrating force 100 yards away into the heather, scoring and scorching the ground for nearly 200 yards and breaking up as it went. Everyone on board must surely have been killed instantly.

David and Hugh Morrison heard the double crash and the

sudden silence of the engines. It was obvious to them that the aeroplane which had just passed over them had flown into high ground. But they could only guess at its position, and in any case they would be able to do very little by themselves. Quickly young Hugh Morrison ran down the sloping hillside to the rough stony track at the bottom, where he had left his motor-bike. He jumped astride the saddle and drove at once into the village of Berriedale to get help. On the way he alerted estate workers, crofters, and shepherds, and in Berriedale he called on the police and the local doctor – the 71-year-old John Kennedy. For the rest of that day a group of volunteers combed the moors and hillsides in the area, but the mist persisted and they could not find the wreckage. Eventually they had to give up as darkness fell.

It was not until one o'clock next day that the search party saw pieces of shattered wreckage glinting below the high shoulder of Eagle's Rock. The flying-boat had broken into a hundred pieces and wreckage was scattered all over the moors. The searchers counted eleven bodies strewn around the main wreckage. Some distance away, lying in the heather in an attitude of repose, was a man with the single thick ring of an air commodore sewn on the sleeve of his uniform. He was the first son of an English sovereign to be killed on active service for more than 500 years.

All 15 of the Sunderland's complement were reported dead. But, miraculously, one man was still alive. One minute he had been sitting in the tail turret of the Sunderland, hemmed in by cloud, and the next thing he knew he was picking himself out of the heather. What had happened was that when the flying-boat struck the ground the tail had broken off and Andrew Jack in the turret had been cushioned as the turret hit the ground. His face and arms had been badly burned and he had been dragged along half-in and half-out of the turret for some distance. Finally he had been thrown out. He knew he had been dragged in this way not because he had any recollection of it but because of the position of the turret and the way the soles of his flying-boots had been torn back at the heel.

His first thought was for the passengers and the rest of the crew, and he started to limp towards the main wreckage. He couldn't find the hull at all, and then he realized that it must have broken up completely. The largest pieces of wreckage were the engines, scattered and smoking amongst the heather. With the aircraft in such minute pieces there was little hope that anyone else could be alive. Peering through the mist and drizzle, he came to the bodies of his companions one by one. All were dead. Then he lost consciousness again.

When he recovered for the second time he found that his eyes were puffy and swollen and everything was hazy and indistinct. He did not realize that he had been burned about the face. But in spite of his difficulty in seeing he realized that he was on a remote hillside and that if no one had seen the plane or heard it crash there would be little hope of rescue. He looked for a path, and eventually stumbled across a narrow shepherd's track alongside a stream. He set off along the track in a downhill direction, taking the easy way.

His uniform was shredded and singed and his torn flying-boots were painful. Soon he got rid of his uniform trousers and boots. He walked and walked by the stream until it got dark. Then he made a bed for himself amongst the bracken and tried to sleep. When he awoke next morning it was a beautiful day, gloriously warm. He had slept right through to midday. His vision was better, and he looked for some point of vantage from which he could search for some cottage or habitation. To his right there was a hill and he started walking towards it, but he made slow progress. He was surprised at his weakness. Walking for 200 yards or so and then resting for a few minutes alternately, he climbed the hill, but it took him a long time and did not seem to give him the view of the surrounding country that he needed. It was late afternoon before he saw a small cottage and turned towards it. Soon he was comfortably installed inside the cottage, drinking a glass of milk and having his burns and abrasions dressed.

Three hours earlier the search party had found the wreck of the Sunderland. It was the 71-year-old Dr John Kennedy who recognized the Duke of Kent, and it was he who took charge of the platinum wrist-watch that the Duke had been wearing. It had stopped 32 minutes after take-off. What could have happened in those 32 minutes? What could have gone wrong? How was it that this highly experienced Sunderland crew, reinforced by their commanding officer – a navigational expert – could have made the sort of mistake that even the rawest beginner was trained to avoid?

Beware of high ground. High ground is a killer. These and phrases like them were dinned into every wartime air force crew. It seemed outside the bounds of credibility that Frank Goyen could have risked a shortcut across such terrain, should have descended amongst the very hills that his briefing and flight plan had been at pains to avoid. It didn't make sense. If he had been a thousand feet or so higher, he might have been suspected of taking the short-cut. Mount Morven, a few miles further inland, was almost the only point above 2,000 feet in Caithness. But at 900 feet and descending – it was the act of a madman. And Goyen was not flying alone. He had a second pilot, a navigator, and his commanding officer on board. They must all have been aware of the course the plane was flying. Yet none of them apparently raised any objection to it. Indeed they must all have approved it. Clearly there had to be some other explanation. Yet a full court of enquiry failed to find one.

Before the departure from Invergordon, said the court, the correct procedure for briefing the captain on the exact route to be followed and for providing full information about the weather conditions had been complied with. Local conditions, it was admitted, had not been good at the time of take-off, but the general indications had shown that there was likely to be an improvement. The pilot was a man of long experience on Sunderlands and of exceptional ability. Yet the pilot, concluded the court, was to blame.

The court's main findings were that the accident occurred because the aircraft was flown on a track other than that indicated in the flight plan given to the pilot, and at too low an altitude to clear the rising ground on the track. This of course was self-evident; there was no attempt to explain how such a tragic aberration could have occurred. The court went on to express the opinion that the responsibility for the mistake in airmanship lay with the captain. This, perhaps, was more controversial. In cases like these, when an aircaft is lost through some baffling mistake that on the face of it looks like an error of judgement, and none of the responsible officers survives, the captain almost invariably gets the blame. But surely in this instance such a facile solution is insupportable, and at this distance ought to be rejected.

What about the weather? Does the court's statement that local weather conditions were 'not good' hide a belief that the Sunderland should never have flown that day? Could this even be yet another case of a wartime VIP's impatience overruling the better judgement of the crew? Those who knew Frank Goyen will scoff at this one. One of the men who served on 228 Squadron with him wrote afterwards that Goyen had a pet retort whenever it was suggested that he might be persuaded to undertake some near-lunatic task. 'I wouldn't do that,' he would say, 'if the King told me.' Neither would he have done it for a Duke. Although the phrase was said in fun, no one doubted that he was capable of meaning it. Whatever may have been hinted to the contrary, no one who knew anything about the Duke himself or Frank Goyen would countenance such a solution. And in any case the court findings gave the opinion that the weather encountered should have presented no difficulties to an experienced pilot.

Could it be that the Duke expressed a desire to overfly some point on land in which he was interested? This might be plausible if there was any such point in the danger area. But there wasn't. The only likely point might have been Dunrobin Castle, and they had passed this 20 miles back and before they entered cloud. Such a request anyway would have been quite uncharacteristic of the Duke and more than ever unlikely in the conditions of a heavy load and low cloud.

The only other published finding of the court was that an examination of the propellers had shown that all four engines were under power when the aircraft struck the ground. The inference was that there was nothing wrong with the aeroplane. Certainly the evidence is all against any engine or structural failure; had there been any such failure, Andrew Jack would surely have known of it. The evidence on the serviceability of instruments and controls, though, is not so conclusive. In the fragmentation that followed the crash, one cannot feel that the fact that the engines were working normally offers any absolute proof of the state of the aircraft.

Having regard to all the circumstantial evidence, and with due respect to the official court of enquiry, one is bound to demur at the placing of blame in this case on one man. In that 32 minutes between take-off and oblivion, something went wrong in Sunderland W4026, something that confused or misled the crew, something that would have confused or misled any other crew at that particular moment in the history of flight. That much one feels is certain. What it was will almost as certainly remain a mystery.

13

THE INFERNAL MACHINE

IT was five minutes past six on the evening of 13th April 1950 as Captain Ian Harvey touched down in his BEA Viking at Northolt at the end of a round trip to Paris. It had not been a pleasant flight. In the unpressurized Viking, a civil version of the wartime Wellington, it was impossible to climb above the weather. The airliner had flown through heavy squalls, thunderstorms and turbulence, and icing had been severe. For safety reasons, and to make the trip less uncomfortable for his passengers, Captain Harvey had flown round the storms when he could, but it had been impossible to avoid a severe buffeting. Throughout the trip there had been the risk that the plane might be struck by lightning, an experience that Ian Harvey had had twice before.

Captain Harvey had one more flight ahead of him that evening – a second outward trip to Paris, where he and his crew would stay the night. He was due to take off at a quarter to eight.

Thirty-year-old Ian Harvey, DFC, a wartime bomber pilot, was a slim, pipe-smoking, phlegmatic man from Bristol, ideally suited temperamentally to civil flying. Like the rest of his crew he was beginning, after the round trip to Paris, to look forward to some food. But tonight's trip meant a smart turn-round and left no time for a meal. Within half an hour of landing and clearing his papers he was back in the operations room to file his flight plan for the final trip. There he met another BEA Viking pilot, Captain George Bell.

'What time are you off, George?'

'Fifteen minutes before you, Ian.'

'Night-stopping in Paris?'

'Yes.'

'See you at the Club, then.' Air France operated a club in Paris at which many of the civil crews ate their meals. 'I shall be famished by then. You can order me a nice, fat, juicy steak.' Harvey's mouth watered at the thought. In meat-rationed England, such a thing was a rarity.

Frank Miller, Harvey's first officer, echoed this request. 'Could you make it the same for me?'

Captain Bell grinned. 'All right. But don't be late or it'll spoil. See you there.'

Harvey called on the Met. man for the latest news of the weather. He was relieved to hear that the storm was moving north-eastwards and would soon be centred over Belgium. They might meet traces of it further south over the Channel, but the squalls and showers would die out over land as night fell.

It was still daylight as Captain Harvey walked from the low prefabricated control buildings at Northolt across the tarmac to the waiting aircraft. The Viking in which he had completed his round trip had already taken off on another service. He was taking over a machine that he had flown many times before and knew well, Viking G-AIVL – Victor Love, in phonetic parlance. But he stepped up the ladder at the rear of the plane with mixed feelings. On the two occasions when he had been struck by lightning, it had been in this aircraft.

Being struck by lightning in the air was an alarming experience, more so perhaps than the actual danger warranted. It was very rare for a plane to suffer more than superficial damage on these occasions. The characteristic portents were a bright flash, a loud explosion, and an acrid smell of burning. Because of the earthing of all metal parts of the aircraft's structure, lightning normally discharged itself into the atmosphere. The fire risk was negligible But sometimes there was damage to external fittings, such as aerials, and occasionally to control surfaces and the aircraft's structure.

Captain Harvey taxied out to the runway shortly before a quarter to eight. The 27 seats of the Viking were all full. One passenger was carrying a three-month-old child. In addition to his first officer, 'Dusty' Miller, with whom he had flown many times, Captain Harvey's crew consisted of Radio Officer Mike Holmes and Stewardess Sue Cramsie. Holmes, a wartime marine operator in the Merchant Navy, was on loan from BOAC and had been flying with Harvey for about a fortnight. Sue Cramsie, cover girl of a contemporary BEA magazine, had had four years as a nurse with the British Red Cross during the war. She had not been on the earlier trip to Paris with Harvey that day, and she nearly missed the evening trip as well. When she reported for duty, one of the stewards, anxious to get a night-stop in Paris, had offered to take her place. She would have accepted, but friends had dropped her at the airport by car and she had no transport to get back to her flat in Eastcote. She decided to go through with the trip.

The storms of the day had flushed the air of south-eastern England and left the sky clear and fresh. But there was no moon, and as night fell the darkness became all but absolute, relieved only by a canopy of stars. Below them they could see the lights of the country towns of Kent and Sussex, and the bright splashes of the coastal towns beyond. Then there was nothing but the unrelieved blackness of the English Channel.

Directly ahead, as they crossed the coast at 5,000 feet near Hastings, they could just discern a dark horizon of storm cloud, stretching across their track like a mountain range. In the diffused light from the stars it was impossible to distinguish a definite outline, but occasional flashes of lightning etched the giant anvil-shapes into momentary relief. Captain Harvey spoke to his radio officer. He did not use the inter-com system, since by raising his voice he could make himself heard.

'What's the radio like? Any static?'

'Not much – except for those lightning flashes. I can hear those all right.'

'Keep the trailing aerial in, will you, Mike? We don't want to attract anything.'

'Roger.'

The light from the stars was now obscured by high cloud. But the cockpit lights were off and Harvey had acclimatized his night vision. The bank of storm cloud lay directly ahead – perhaps not more than a mile away.

'I don't want to fly through this, Dusty. Can you see a way round it?'

First Officer Miller formed his hand into a pointer and indicated an oblique direction to port, about half-left. 'I think it's only the residue of the day's storm. The worst of it is a long way away now. I think this stuff peters out over there.'

'I'll give George a call. He must be through it by now.'

Harvey called Bell on the voice radio, and Bell confirmed Dusty Miller's opinion. Harvey disconnected the automatic pilot and altered course to port. No doubt he could fly through this remnant of the storm quite safely, but it would be bumpy, and in any case the safer course was to fly round it, since he could not climb over it. He decided to hand-fly the plane until they were clear of the storm.

In the cabin, Sue Cramsie was handing out the cardboard supper boxes to the passengers, and serving coffee. When everyone was served she returned to the rear of the plane and sat down on the tip-up seat in the pantry, just behind the rearmost seat on the starboard side, immediately opposite the main door. She rarely got a chance to sit down on this evening flight to Paris, but tonight the passengers were an abstemious and undemanding crowd. She would give them time to finish their coffee, and then she would open the bar box at the extreme rear of the cabin and take round a selection of duty-free cigarettes.

Suddenly she was aware of a faint, peculiar smell which quickly gained in pungency. One got used to the typical smells of aeroplanes, and this was not like any that she could remember. It was completely unfamiliar. It wasn't oil, or petrol, or the heaters, or dope, or coffee, or anything that she could identify. Possibly one of the passengers was smoking some unusually exotic cigarette. But the smell transmitted a single word to her mind. Acid. It was harsh and sharp, and she was aware of a stinging sensation at the back of her nose and throat.

She left her seat in the pantry and moved into the cabin to investigate. She stood looking down the gangway between the seats, staring inquisitively at the passengers. No one seemed to be doing anything unusual. And out here the acid smell seemed to have gone.

It occurred to her to go forward and tell the crew. But if one troubled the crew every time one smelt something a bit odd in an aircraft, one wouldn't last long as an air hostess. In any case she had probably been mistaken. It could be nothing to worry about.

Now that she was up, she might as well do the cigarette sales. She turned round to fetch the tray. But the passengers on Viking Victor Love that evening were not destined to get their duty-free cigarettes. As Sue Cramsie moved towards the rear of the cabin, she was shocked by a violent explosion to her left. She had time to raise her left arm to protect her face, she felt the deafening impact of the explosion, and she sensed that she had been blown off her feet. Then she lost consciousness.

In the cockpit, Captain Harvey was just deciding that he must be clear of the danger from the storm cloud, and that the time had come to turn back on course, when there was a flash and a muffled explosion, followed by a fierce rush of air and an acrid smell of burning. In spite of all their precautions they had evidently been struck by lightning.

He felt the controls stiffening under his grip. There seemed to be some external pressure on the control column, pulling it back, lifting the nose into a steep climbing attitude. It took all his strength to push against the control column and check the upward motion. Light from the cabin was streaming into the cockpit and his night vision was destroyed. For a moment he was almost blinded. He didn't want the passengers staring in, and he shouted at Holmes, the radio officer.

'Shut the door, Mike!'

'I can't. It's blown in! It hit me on the head!'

The lightning must have struck them aft. There must be something seriously wrong back there, something to account for the jammed controls and the howling gale that was blowing through the fuselage.

'Dusty! Go back and see what's happened in the cabin.'

As Miller hurried back into the fuselage, Harvey began a

systematic check of the controls. He tried a fairly gentle turn, first to port and then to starboard. The aircraft responded normally. But all the time there was this tendency to climb, this backward pressure on the control column. When he tested the elevators he found as he suspected that they were locked almost solid. There was not more than an inch of movement, and some unknown force was tugging them back.

He kicked at the rudder pedals, and they slammed loose against the stop, then flapped brokenly. He had no rudder control at all. He tested the rudder and elevator trimmer wheels and they spun loose through his finger-tips. There was no control there either.

His muscles ached with the struggle to hold the plane straight and level. He couldn't keep this up much longer. He tried wedging his knee between the seat and the control column, and found with relief that it enabled him to relax his shoulder muscles. He had found a way of keeping the plane on an even keel.

What about undercarriage and flaps? What would be the problems of getting the aircraft down? He selected undercarriage down, and was relieved to see the three green warning lights, port, starboard, and tail-wheel, flash on. He extended the flaps about a third, and they operated normally. This was comforting, but he was still faced with the problem of flying the aircraft with damaged controls and of manoeuvring it to get it down. He judged they were rather less than half-way across the Channel. It might be safer to turn back to Northolt.

Miller came back into the cockpit and shouted in his ear. 'It's a shambles back there. There's a gaping hole on each side of the fuselage and the door's blown open. Sue's badly hurt. She's covered in blood and unconscious. It's a wonder she wasn't blown out. The passengers have pulled her clear and covered her with blankets. They're all behaving magnificently – calm as anything and offering to help. Everyone else is OK.'

'That settles it. We'll go back to Northolt.'

The decision was instinctive rather than reasoned, the normal instinct to get back home when in trouble, but in the back of Harvey's mind was the knowledge that the BEA organization was inevitably better at the home base than in Paris, where the French would be helpful but where the language difficulty might complicate matters in an emergency. Keeping up the forward pressure on the control column and using his aileron control, he turned gently through 180 degrees until the Viking was flying back towards the English coast. Then he called to Mike Holmes.

'Get through to Uxbridge, Mike.' Uxbridge was the traffic control for Continental flights. 'Tell them we've been struck by lightning and that the aircraft was damaged, the air hostess is badly injured, and we're returning to Northolt. Ask them to have an ambulance ready to meet us.'

While Holmes tried to clear his message, Harvey explained and demonstrated to Miller how he had managed to retain control. Holmes was unable to make contact at first, so he prefaced his message with the special urgency signal 'XXX'. At once all other air operators stopped transmitting and left the ether clear.

When he heard from Miller the extent of the damage in the rear, Captain Harvey realized that probably the gravest danger was that the aircraft would break up in the air. The hole on the starboard side, where the lavatory had been, was eight feet high and four feet wide, big enough for two people to walk through side by side. There was a similar and only slightly smaller hole on the port side, through which Sue Cramsie had nearly been thrown. The effect of these two enormous holes must be to put an increasing strain on the tail of the aircraft and weaken the structure at a critical point. The influx of air would also tend to increase the weight and drag in the tail.

Harvey decided to reduce speed from 170 knots to 135, to ease the strain on the fuselage structure. Then, after satisfying himself that Miller had the feel of the jammed controls and could safely take over, he went back into the cabin to see the damage for himself and reassure the passengers. When he saw the twisted, tangled disintegration in the tail he marvelled that the plane was still flying. All the control rods from rudder and elevators must be bent or broken. It was astonishing that there was any response from the elevators at all. He had never seen a lightning strike like it. It was worse than any flak damage he had seen during the war.

It was unsafe to go too near either of the gaping holes, and impossible to make any detailed check of the damage, but he saw enough to realize that the whole tail section was now supported by nothing more than two narrow strips of metal, one along the roofing at the top and the other under the floor, where the fuselage was specially strengthened to bear the weight of the aft freight compartment. This was probably what had saved them so far. It occurred to Harvey, too, that it was quite possibly a good thing that the rudder had jammed. Lateral movement by means of the rudder would put such a strain on the thread of metal supporting the tail that it might well snap.

He bent down to speak to Sue Cramsie, who was beginning to revive. In a half-bemused state she had reached across with her right hand to feel the injuries to her left arm and then hastily withdrawn it. Half her arm between elbow and shoulder was a clammy, cloying gash of torn skin and blood. In that brief moment of contact her fingers had touched the bone.

She looked up to see Harvey gazing down at her.

'Shall we be all right?'

'Of course. We're going back to Northolt. We'll be there in a few minutes. There'll be an ambulance waiting for us.'

'I'm so cold.' Now that she was aware of her injuries, the shock and pain were redoubled. They were so intense that she knew she would faint again. 'Can you get me some morphine?'

Harvey turned involuntarily towards the exit door, in which the first-aid kit was housed. The door was hanging on its hinges in space. It was quite impossible to reach the box.

'I can't reach it.'

She accepted this for the moment, and then when Captain Harvey, after spreading another blanket over her, went forward to the cockpit, she wondered how it could be that he couldn't reach the box. Then she lapsed again into unconsciousness.

Back in the cockpit, Harvey called George Bell on the voice radio. 'Fox Mike from Victor Love. We've been struck by lightning. We're returning to Northolt.'

Captain Bell's reaction was to roar with laughter. 'Careless! You should have looked where you were going! What, no steak?' Harvey found Bell's callous amusement surprisingly heartening.

He called Northolt. They were just crossing the coast when Northolt answered. He repeated the urgency message already sent by the radio officer, and Northolt acknowledged it. 'You are clear to make a direct approach and land,' they told him. All other aircraft movements in the zone had been frozen or suspended and the skies below 4,000 feet were empty. Harvey crossed southern England at 2,500 feet on a long, straight approach into Northolt, and eight miles from the airfield he began his final let-down. Soon the Viking was floating in over Harrow-on-the-Hill.

'Get back into the cabin for the landing,' he told Mike Holmes. 'Brace yourself against the main spar.' Holmes went back and saw to it that the passengers were strapped in before taking up his crash position on the cabin floor. His gay, infectious good-humour gave confidence to the passengers. They had never shown any tendency to panic, but whatever their secret fears might be, it was clear to them that here was a man who had no doubts at all.

In the cockpit, Harvey and Miller strapped themselves in, using their full safety harness, their shoulders braced tightly back against the uprights of their seats. Harvey's intention was to trust to his ailerons for turning and to use his throttles to control his height. It would be just before touch-down that the lack of elevator control would be most critical. When it came to holding off for the landing, he would have to try to get some movement from the elevators.

'If things don't work out on the final approach,' he told Miller, 'I'll give you a shout. I want you to use your weight on the control column to help me push it forward or pull it back – whichever is needed.'

'OK, Ian.'

They could see the runway lights now, and Harvey put his undercarriage down, confirmed that it was locked, then put down 20 degrees of flap. He was losing height smoothly and easily by manipulating the throttles, keeping his knee jammed between the seat and the control column. But it was axiomatic that when he took power off the engines the plane would sink. He had to judge the drop correctly over the last half-mile.

At 500 feet he put down 60 degrees of flap. It was an odd feeling having no elevator control, but keeping one hand on the throttle levers he coaxed the plane forward in an accurate glide. He was down to 25 feet, right over the start of the runway, when the nose started to drop.

'Help me to hold off!' The reaction was instant and instinctive as both men hauled back on the jammed controls to lift the nose and prevent the aircraft from plunging into the runway. The combined weight of both men achieved its purpose – too well. They had over-corrected. Released from the forward pressure that Harvey had hitherto been forced to apply, and with both pilots pulling back, the jammed elevators lifted the Viking into a steep climb. With the speed already down to 100 knots for the landing, it could only be seconds before the plane stalled and spun in.

The only hope was to pour on the power. Captain Harvey rammed both throttles wide open. The effect was startling. The increase in power only made the nose-up attitude worse. Their speed was still dropping off. Miller, watching the instruments, saw the indicated air speed flicker down to 90 knots – just above the stall. Both men pushed forward on the control column with all their might to bring the nose of the Viking down, but the speed continued to drop. The needle flickered down to 80 knots, the measured stalling speed. Miller daren't look any more, and he turned his head away.

It was no good throttling back. They needed the power. Under the Herculean pressure applied by both pilots the nose at last began to fall. But what if this pressure should snap the buckled control rods in the tail? It would be too late then to restore an even trim. The Viking would fall out of control.

Had Captain Harvey known the full extent of the damage in the tail it is likely that he would have risked a crash-landing rather than subject the aircraft, and the damaged controls, to the strain of an overshoot and a second attempt at landing. But such knowledge could only come from a full examination on the ground. What he had done in opening the throttles and going round again had been the pilot's instinctive reaction to one of the most dangerous moments in flying – the sudden loss of air speed on the approach to land, threatening the stall. Yet somehow Viking Victor Love stayed in one piece as Harvey made his second circuit. The inch or so of elevator control remained. With the use of aileron he was able to bank round the airfield and line up again on the runway.

He had learnt a lot on that first approach. This time he used less flap, made a more shallow approach, and timed his arrival over the runway to a nicety. Keeping the nose at the right attitude without any assistance from Miller, he managed by judicious use of the throttles to lower the plane gently towards the runway and flatten out. By a miracle, they were safely down. Miller shouted at him triumphantly. 'Skipper, that's the best landing you've done all day!'

Charging down the runway after them came the fire engines and the ambulance. Only the ambulance was needed. Harvey taxied to the tarmac, and Sue Cramsie, her back pitted with small pieces of metal, her wrist and forearm gashed and torn, her upper arm almost severed, was taken to hospital. Fortunately the bone was intact and she subsequently made a complete recovery.

In the light of the arc-lamps on the tarmac, Harvey and Miller stepped down from the Viking and walked back to the tail, surveying the damage. It was Miller who first voiced their combined suspicions.

'We've never been struck by lightning!'

'What do you mean?'

'Lightning couldn't do that! Look! All the jagged metal is

twisted outwards. It's the same everywhere. The explosion must have come from inside.'

'Lightning Blasts Airliner!' said the headlines next morning. But already an investigation had been begun which was to prove that the elements were not responsible for the violence done to Viking Victor Love. The first suspect was the disinfectant fluid used in the lavatory, and before midnight that night BEA had signalled all stations to change this fluid in all their aircraft. But inflammability tests proved that neither the fluid nor its vapour would burn or explode. And meanwhile it had become clear that the centre of the explosion had been at the extreme rear of the toilet compartment, the farthest point away from the Elsan toilet.

There followed a most painstaking technical and scientific enquiry into the cause of the explosion. But nothing was found which could be indentified as part of an 'infernal machine', as the board of enquiry termed it, and the only extraneous object in the wreckage was a used match, burnt almost its entire length. The likely location of the explosion was pinned down to a dirty-towel box immediately behind the toilet door. The box itself was never found, but metal fragments of a similar light alloy, showing definite explosion characteristics, had penetrated the wash-basin and other parts of the toilet compartment. 'It was almost certainly this box,' said the report of the enquiry, 'in which the explosive object was planted and detonated.'

Who planned to destroy Viking Victor Love on that April evening in 1950, with all its passengers and crew?

Within 24 hours of the accident, Scotland Yard had moved in. Already the passengers had scattered. Most of them, showing commendable fortitude, had transferred to a later plane that same night and completed their journeys, to Paris and beyond, before foul play was suspected. Others travelled next day. All were under the impression that their flight had been interrupted by a natural phenomenon. All, that is, unless the bomb was planted by one of the passengers.

This seems the most unlikely of all explanations. Yet in spite of exhaustive enquiries, Scotland Yard were unable to solve the mystery of who planted the bomb, or who stood to gain by the Viking's destruction. Or if they solved it, they were unable to make an arrest. Various theories were put forward. A ruthless smuggling ring, it was said, had decided that one of their contacts was no longer reliable and must be liquidated. What more natural than that they should place their explosive in the receptacle which, perhaps, had hidden their illicit consignments? Straightforward murder, murder for insurance money, murder for political ends, murder for spite – all these possibilities were canvassed and investigated. Still no arrest was made.

Could it be that in this case the most unlikely solution at first glance is in fact the most feasible? A human being desperate enough to destroy himself does not always baulk at the mass destruction of others. History provides notorious examples. And surely a macabre attempt at suicide fits the circumstances best.

Let us suppose that one of the passengers, driven to suicide, determines to disguise it so that his dependants can benefit from his insurances. In moments of despair it has often occurred to him that to be killed in a fatal air crash would be the answer to his agony. It is a short step from here to the precipitation of such a crash.

He carries the necessary explosive charge with him. It is simple, perhaps crude, involving nothing more than the explosive, a detonator, and a length of fuse. He waits until the plane is over the sea. Then he makes his way back to the toilet. He puts an explosive charge in the dirty-towel box, strikes a match, and lights the fuse. To make quite sure it is properly alight, he lets the match burn right down. It burns so low that it scorches his fingers. Dropping the match to the floor, he returns to his seat.

In the pantry, the air hostess smells a peculiar burning aroma. Then comes the explosion. Everyone on board is more than a little frightened, and the behaviour of the guilty passenger, unnaturally calm or oddly neurotic, passes unnoticed. Scotland Yard develop their suspicions of two or three people, perhaps of one person only, but the passengers have scattered and the evidence available is meagre. They can find no damning clue to support a prosecution, such as the date and place of the purchase, by an identified passenger, of a quantity of explosive. The case drags on.

The file at Scotland Yard, perhaps, is still open. Even now, over fifteen years later, there may be someone who fears arrest, an arrest that may yet be made.

People who tamper with aircraft always leave tracks.

14

BILL LANCASTER

'THE beasts of the field fight to the death for a female. The birds of the air do likewise. This war hero is not the type of man to stand aside like a coward when a man takes from him the woman he loves.'

With these somewhat highly coloured words the State Attorney in the courtroom at Miami, Florida, brought his closing speech for the prosecution to a telling climax. The man in the dock was a world-famous flier whose fortunes had reached their nadir. In what looked like a routine love triangle he was accused of murdering the other man. The evidence against him was damning.

Born in 1898 at King's Norton, Birmingham, Bill Lancaster had been educated at Ardingly and Stafford Colleges and had emigrated to Australia on leaving school, a few months before the outbreak of war. In 1916 he enlisted in the Australian Army, and he served in the Middle East and in France before joining the Australian Flying Corps to train as a pilot. Later he transferred to the newly-formed RAF. After the war he had a spell on the reserve before rejoining the peacetime Royal Air Force. In the meantime, at the age of 21, he had married. In addition to his skill as a pilot he was a first-class amateur steeplechase rider and amateur boxer; in 1924 he won the amateur broncho-riding competition at Wembley Rodeo, and he captained the RAF boxing team in 1925.

In 1926, after five years on the active list, Bill Lancaster's engagement in the RAF ended. He did not settle easily to a civilian career, and when in the following year a chance came his way to attempt a flight to Australia in a light plane he decided to take it. The project was not just another stunt but was intended as a reliability test for the newly designed Avro

Avian and its 80 horse-power Cirrus engine. Lancaster's co-pilot and companion on the flight was to be an Australian woman flier named Mrs 'Chubbie' Miller, dark, petite and self-reliant, wife of a prominent Australian journalist. Mrs Miller had met Lancaster three weeks before the flight and had begged to be allowed to accompany him.

Lancaster and Mrs Miller left Croydon on 14th October 1927. They were not attempting anything spectacular in the way of speed and they were not trying to beat any records. Just to fly from England to Australia in a light plane would be a record in itself. It was a protracted flight, made all the longer by delays for bad weather and unserviceability and eventually by a crash-landing on an island off Sumatra after a sudden failure in the fuel system. While they were held up, in February 1928, Bert Hinkler passed them and became the first man to fly a light plane – it was another Avro Avian – to Australia. It was not until 19th March 1928, more than five months after leaving Croydon, that they finally reached Port Darwin.

Inevitably, through many dangers and vicissitudes that were interspersed with long hours of boredom and loneliness, Bill Lancaster and Chubbie Miller were thrown closer and closer together. Just as inevitably, they fell in love.

After the Australian flight, Lancaster was engaged as demonstration pilot for the Cirrus engine, and in March 1929 he began an American tour designed to publicize British light planes and engines. The plane he flew was again an Avro Avian. He was accompanied on this tour by his wife and by Lady Heath, who in the previous year had made the first solo flight from South Africa to England. Lancaster by this time had two young children, both girls, and it seemed that he and Chubbie Miller were doing their best to put their experience behind them. But eventually the mutual attraction proved too strong. Mrs Miller took a bungalow in Miami, and Mrs Lancaster returned to England.

Times were difficult for a free-lance flier in America after the 1929 Wall Street crash, and things were no better in England. Lancaster and Mrs Miller stayed in Miami, while

Lancaster picked up what flying work he could. Mrs Miller's divorce came through in 1931, but there seemed no prospect that Lancaster's wife would divorce him, so the couple were unable to marry, although they intended to do so when they could. Several times they had discussed the problem with a Miami attorney called Huston, but there was nothing they could do. Lancaster idolized Chubbie Miller; but within two or three years her feelings for him underwent a change. She remained intensely fond of him, admired him tremendously and trusted him implicitly; but she was no longer in love with him.

To find work Bill Lancaster was forced to go further and further afield. Sometimes he was away for weeks at a time. Essential bills sometimes went unpaid. In an attempt to earn some money, Chubbie Miller accepted the offer of a young airman/author to 'ghost' her life story. He had written several books, he told her, and he thought this one, with its background of flying and adventure, would sell. His name was Haden Clarke.

On 6th March 1932 Bill Lancaster went away again on an air tour, leaving Chubbie Miller and Haden Clarke to get on with the book. Clarke was a weak character, neurotic and unstable, already involved – unknown to the others – in a mesh of lies and deceit. Chubbie Miller, herself a victim once before of propinquity, felt sorry for him. For a time at least, with Bill Lancaster away, her sympathetic feeling for Clarke came very near to love. When he pleaded with her to marry him, she agreed.

Nightly from his stopping points all over the southern States Bill Lancaster rang to talk to Chubbie. She could not dissemble, and he detected a change in her tone of addressing him. Puzzled by her behaviour and suffering tortures of doubt, he became ill with nervous worry. Chubbie Miller was determined not to deceive him for longer than was absolutely necessary, and when she became sure of her feelings she wrote to tell him that she had fallen in love with Clarke and that they were to be married. Clarke, too, wrote to him at her instigation to explain.

Lancaster's reaction to these letters was one of deep

depression and bewilderment. The main thing was to stop Chubbie doing anything irrevocable before he could return and talk things over with her. He cabled her at once from St Louis. 'Congratulations to you both, but wait until I return so that I can be best man.' This attempt at gaiety and sportsmanship hid a state of acute mental agony. For many hours he contemplated suicide. Before he left St Louis to fly back to Miami, he bought a gun.

Lancaster reached Miami on 13th April. He had been away five weeks. He arrived in a depressed state, but he had no reproaches for Chubbie. The triangle was completed that evening as the three of them sat at dinner at the bungalow. Soon the conversation came dangerously to the point.[1] 'Haden, old man,' said Lancaster, 'you betrayed the trust I reposed in you. It wasn't the act of a gentleman.'

'I resent that.'

The highly emotional Clarke seemed almost more upset than Lancaster. Lancaster was trained to hide his feelings and to keep his temper. Clarke's make-up was entirely different. There was a stormy scene in which harsh words were spoken. Eventually Lancaster extracted a promise from the two of them that, to give Chubbie time to think things out and be sure of what she wanted, they would not marry for at least a month. After this Clarke became self-condemnatory. 'I guess you're right, Bill,' he said. 'I betrayed you. I let you down.' Chubbie Miller, too, was remorseful. 'I can't bear hurting Bill like this,' she said to Clarke when Lancaster left them for a moment, 'I wish we could end it all.' Clarke responded to this with surprising calm. 'Yes,' he said, 'it would be a solution. I wish we could.'

Lancaster and Clarke made up two single beds on the verandah of the bungalow. When Chubbie Miller went to her room, Clarke asked her to lock her door. 'Remember that you promised to marry me,' he whispered. 'I don't want Bill creeping in to talk you out of it.' She did as he asked.

It was late when the two men at last turned in. 'Bill,' said

[1] The ensuing dialogue is quoted from the evidence given by Lancaster and Mrs Miller before and during the subsequent trial.

Clarke before they went to sleep, 'you're the whitest man I know.' They were the last words he spoke.

During the night Chubbie Miller heard a pounding on her door. It was Lancaster. Haden Clarke, he said, had shot himself. He was bleeding from a wound in the temple but was alive and was trying to say something. Lancaster sent at once for a doctor and an attorney – the man named Huston whom he had consulted over his divorce. It didn't take the doctor long to make his diagnosis: Clarke was dead. The revolver that Lancaster had bought in St Louis lay by his side. Lancaster, afraid that suspicion might fall on him, begged Huston to say that the revolver was his, but Huston declined to do so and sent for the police. The police, however, were quite satisfied. They found two suicide notes, typewritten and signed by Clarke, one to his mother, the other to Chubbie Miller. 'Chubbie,' it said, 'I can't stand the economic pressure. Will you help to sustain my mother in her deep grief?' In the note to his mother he begged her forgiveness.

A few days later, after Haden Clarke's funeral, it seemed that a distressing story had reached its tragic end. But a week after his death the case was re-opened. A handwriting expert had testified that the signature on the suicide notes was a forgery; the whole thing could have been concocted by someone else. Suspicion at once fell on Lancaster. He and Clarke had been rivals for the love of the same woman. He had quarrelled with Clarke that evening. It had been his gun. On 21st April he and Mrs Miller were detained pending an investigation. Mrs Miller was released almost at once, but on 2nd May the Florida State Attorney formally charged Lancaster with murder, refusing bail. Six days later Lancaster was indicted before a grand jury on a charge of murder in the first degree, for which the penalty was the electric chair.

For three months Lancaster languished in jail awaiting trial. 'I am absolutely innocent,' he declared. 'At the right and proper time an explanation will be made.' But he did not deny forging the suicide notes. Chubbie Miller stood by him. 'I am convinced that Captain Lancaster will be fully and honourably exonerated,' she said. 'My faith in him is unshaken. I am certain Haden's death was due to suicide.'

But the case against Lancaster looked overwhelming. The authorities had taken possession of his private diary, to which he had committed his innermost thoughts. It spoke again and again of his love for Chubbie, of his distrust of Clarke, and of the heartache and torture he was undergoing. 'My mental agony is hell,' he had written. 'I am determined to have it out with Clarke when I get back.' All this would be produced in court. Motive and opportunity were clear, Mrs Miller would be obliged to testify that the two men had quarrelled, and evidence would be brought by the prosecution that Lancaster had bought the gun at St Louis immediately after receipt of the letters from Mrs Miller and Clarke and then hurried back to Miami. All the prosecution needed was firm evidence on Lancaster's state of mind, and this they had in the diary. Even Lancaster's own defence counsel, James Carson, engaged from a distance by Lancaster's father, at first refused the brief. 'I'm not interested,' he said, 'the man's as guilty as hell.'

But once he had been persuaded to meet Lancaster, Carson changed his mind. Lancaster's story was that he and Clarke had retired to bed on the verandah and chatted amicably for some time. About one o'clock they had gone to sleep. Lancaster had been awakened by a shot. He had got up and found Clarke on the edge of his bed, shot through the head, a smoking pistol in his hand. He had rushed into the bungalow to wake Mrs Miller and to call a doctor. Then, aware that his own gun was missing and that Clarke must have used it, and fearing that suspicion might fall on him, he had typed the suicide notes and thrust them in front of Clarke. Clarke had been too weak to sign them, and after he died and before the police came Lancaster had hurriedly signed them himself. He would never have done it if there had been the slightest doubt in his mind that Clarke had committed suicide.

It was a plausible story and Carson wanted to believe it, but he doubted if it would stand up against a well-organized and relentless prosecuting counsel. He was considerably helped towards belief, however, when he began to investigate the background of Haden Clarke. He found that Clarke was not only married already but had a bigamous marriage to his credit as well. He was a sick man. He was a drug addict. He had vastly exaggerated his writing ability and had never written any books; it was extremely doubtful whether he would ever have been able to produce a book on Mrs Miller or anyone else. Yet with all his deceits it seemed that he was fundamentally decent and was genuinely in love with Mrs Miller, and remorseful over his behaviour towards Lancaster. He had spoken before of suicide. Now, in a mood of self-denigration, with his deceits catching up with him, and aware perhaps that Lancaster was much the stronger character and that Chubbie Miller's infatuation would not long survive Lancaster's return, he had decided on impulse to kill himself. He had not made a very good job of it, but the damage had been enough.

The two sides of the case, prosecution and defence, were thus clear-cut. For the prosecution it was a crime of passion. For the defence it was a case of suicide. To support their thesis the defence asked for an autopsy, holding that an examination of the skull would confirm the theory of suicide.

On Friday, 12th August 1932, in the middle of the trial, the skull of the dead man was taken from a cardboard box and shown to the jury. Women spectators paled and fainted or left the court, but Lancaster leaned forward to see it, showing no horror. Members of the jury passed the exhibit from one to another. Meanwhile medical evidence was being brought by a pathologist and a specialist in ballistics that the bullet hole and the blackened marks on the temple showed that the revolver must have been held close against the victim's head, which meant that the wound was probably self-inflicted. A murderer, argued defence counsel, would hold the gun an inch or two away for fear of disturbing the victim.

This rather circumstantial quasi-scientific evidence may or may not have impressed the jury. Certainly when prosecuting counsel delivered his final speech, several of Lancaster's actions were still not convincingly explained and the theory of a crime of passion looked irresistible. There were, however, three intangible factors which influenced the jury. One was Lancaster's own demeanour. Second was the absolute confidence retained in him by Mrs Miller. Third was the very weapon with which prosecuting counsel had sought to establish his case beyond all doubt – the diary. 'It has been my privilege,' said the judge, 'to see into the depths of a man's soul through his private diary, which was never intended for anyone's eyes but his own, and in all my experience – which has been broad – I have never met a more honourable man than Captain Lancaster.' This indeed was the impression of the entire court. In spite of evidence against him which seemed overwhelming, Bill Lancaster was found not guilty of murder and discharged.

After the glare of publicity that had surrounded him during the trial, and the inevitable popular impression that he had probably committed the crime and done well to get away with it, Bill Lancaster was faced with what seemed an insoluble problem of rehabilitation. Certainly it seemed impossible in America. Eventually he returned to England and tried to build his life afresh there. But there were several strands of it that could not be started again from scratch. His marriage was one of them, his relationship with Chubbie Miller was another. The only friends he had to turn to were his parents; they had financed his defence in the trial, and they now offered him a home and some prospect of financial help until he re-established his reputation as a flier. For their part they asked only one thing in return – that he make a conscientious attempt to be reconciled with his wife.

Getting a job in aviation had become, for Lancaster, just as difficult in England as it had been in America. By the end of that year – 1932 – he had decided that his only hope of rehabilitation lay in making a record flight. Even this depended on financial backing from his father. He chose to attack the record from England to the Cape, a record formerly held by Jim Mollison but recently broken by Amy Johnson. She had completed the flight to Capetown in November 1932 in 4 days, 6 hours, 54 minutes, beating Mollison's record by about 10½ hours.

The plane that Lancaster bought to attack this record was of a type he knew better than any other – an Avro Avian with a Cirrus engine. This one had formerly been owned by Sir Charles Kingsford-Smith, who had named it the *Southern Cross Minor*. 'Smithy' had set out from Australia in it eighteen months earlier in an attempt to beat Mollison's record for the flight to England. He had been subject to many delays and the attempt had failed, but he had completed the course. A biplane with a very low structure weight and a great load-carrying capacity for its size, it was a well-tried machine, designed originally by Roy Chadwick, chief designer and engineer to A. V. Roe. It carried 115 gallons of fuel and had an endurance of approximately 14 hours. At a cruising speed of 95 mph this gave a range in still air of just over 1,300 miles.

For the Cape record the plane had one serious drawback: it was nearly 20 mph slower than Amy Johnson's de Havilland Puss Moth. This meant that Lancaster could only succeed by accurate flying and by cutting his time on the ground at the various refuelling points to a minimum. He did not expect to get more than an hour or two's sleep at any of the stops *en route*.

Both Jim Mollison and Amy Johnson had chosen the more direct western route across central Africa, instead of the old eastern route down the Nile. Lancaster intended to follow the same route; he would have no chance of the record otherwise. The first part of the flight would take him due south to Oran in Algeria, a distance of 1,100 miles, which he hoped to make in one hop if the winds were right, crossing France, the Pyrenees and the Mediterranean on the way. The next leg of the flight would take him across the Sahara. Amy Johnson in her Puss Moth had accomplished the 1,400-mile flight to Gao on the Niger in one hop, but Lancaster in his Avro Avian could not attempt this; he would have to make a short stop for refuelling at Reggan, an oasis outpost on the Trans-Saharan motor track 630 miles south of Oran, and then carry on for Gao from there. He did not anticipate any difficulty in finding Gao; all he would have to do would be to hit the Niger and follow it south. The greatest problem of the whole flight, he believed, would be finding Reggan. He expected to arrive there in darkness.

Having crossed the Sahara he would still have nearly 4,000 miles to go, but by his next stop, at Douala on the coast of the French Cameroons, he would feel that the worst of the flight was over. From then on he would have the African coastline as a constant check on his position right the way through to the Cape. It was over this latter half of the route that he hoped to improve on Amy Johnson's time. She had suffered several delays in central and south-west Africa.

The efforts that were made to repair Lancaster's marriage met with no success; it was Chubbie Miller, her love for him restored and strengthened, who helped him plan the flight. Three days before he took off he made a will in her favour. Apart from insurances, his assets amounted to very little.

Early on the morning of Tuesday, 11th April 1933 – it was Easter Week – Lancaster's parents went to Lympne to see him off. Chubbie Miller, too, was there. To beat Amy Johnson's record he had to reach Capetown by midday on Easter Saturday. 'I owe this chance to come back,' he said, 'to my father and mother. My father has committed himself to heavy expenses so that I can make an attempt on the record, and for their sake alone I hope to win through.'

He then referred to his plans for the flight. 'Beyond a few snatches by the side of my machine,' he said, 'sleep will be out of the question for the next few days. My most difficult task will be the location of Reggan. I must land there to refuel before making for Gao in French West Africa.

'I hope to cover 1,150 miles today and reach Oran this evening. After a brief rest I shall begin the flight over the Sahara.

'I want to make it clear that I am attempting this flight at my own risk. I don't expect any efforts to be made to find me if I'm reported missing.' Privately, Lancaster had said that this would be his final attempt to re-establish himself in British aviation: if he failed, he did not wish to come back.

Lancaster had promised his parents that he would not take any undue risks; but the whole conception of the attempt involved a physical and mental pressure that would be bound

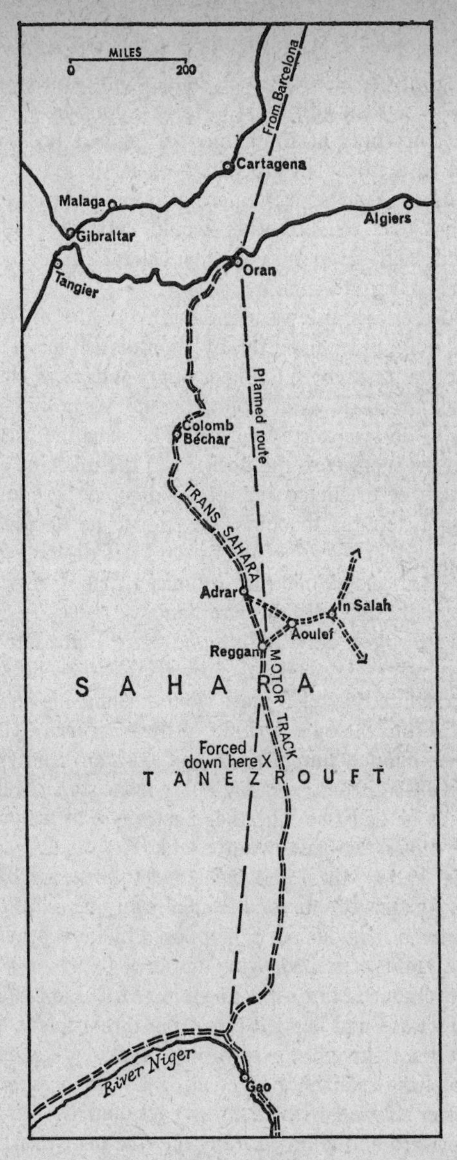
MILES
0
200
From Barcelona
Cartagena
Malaga
Algiers
Gibraltar
Oran
Tangier
Planned route
Colomb Béchar
TRANS SAHARA
Adrar
In Salah
Aoulef
Reggan
MOTOR TRACK
S A H A R A
Forced down here X
T A N E Z R O U F T
River Niger
Gao

8. Oran to Gao

to drive him very near breaking-point. What state was Lancaster in to face such an ordeal? He had done no flying for 12 months. For three of those months he had been in prison, confined to a tiny cell. He had been subjected to a terrible mental strain. Yet the excitement of a record attempt, and the knowledge that rehabilitation would certainly follow if he succeeded, had given him back much of his buoyancy and confidence. Or so it seemed.

One further promise was extracted from him, by his parents and also, quite independently, by Chubbie Miller. The history of long-distance pioneer flying strongly suggested that when a plane was forced down in isolated country, the best thing for the crew to do was to stay with it. The machine was easier to find than the man. Also the position of the machine could to a large extent be predicted; its intended course was known. But once a man left his machine, the course he decided to take would depend on a score of factors and was difficult to forecast or divine. Lancaster therefore promised that if he was forced down he would stay with the aeroplane.

It was still pitch dark at Lympne when Lancaster went out to his aircraft, wearing flying overalls, a wind-cheater, and a brightly coloured woollen scarf wound round his neck. Before he climbed into the open cockpit of the *Southern Cross Minor* his mother handed him a packet of chicken sandwiches, the only food he intended to carry apart from some beef extract. For liquids he had two Thermos flasks, one of water and one of coffee, plus a two-gallon water tank. He was also carrying a sun helmet and a pair of yellow goggles to protect his eyes, and a small cash sum for the payment of fuel *en route*. For talismans he had a religious poem composed by his mother, a silver horseshoe, and two snapshots of Chubbie.

As he climbed into the cockpit, Mrs Lancaster senior stepped forward and handed him a bar of chocolate which she had forgotten to give him. Dawn was breaking, and as the first light revealed the airfield in bleak, shadowy perspective, the little cluster of well-wishers drew back and the Avro Avian moved forward and climbed away into the near-darkness. The time was 5.38 AM.

Interest in Lancaster's flight at this stage was no more than tepid. Record attempts were frequent and they mostly failed. The excitement would begin if Lancaster did well and looked like breaking the record. The first comparison with Amy Johnson's flight would be his arrival time at Oran.

The weather across the Channel and over France was no better than moderate, but Lancaster was able to keep a regular check on his ground speed. He crossed the French coast at Le Havre and continued across France via Tours and Limoges, keeping slightly to port of due south to avoid the high peaks of the Pyrenees. By the time he reached Toulouse he was behind schedule due to an unfavourable wind, and the narrow endurance margin he was allowing himself for the Mediterranean crossing had gone. He was forced to land at Barcelona to refuel; this took him out of his way and entailed nearly two hours' delay. He completed the sea crossing safely but it was nine o'clock that night before he landed at Oran, 15½ hours from Lympne against Amy Johnson's eleven. He was thus 4½ hours behind the record, but he had every hope of making this up and a good deal more over the later stages of the flight.

Refuelling and servicing at Oran took longer than he had reckoned for; he was six hours on the ground there against Amy Johnson's four. That was another two hours to make up. He finally got away from Oran at three o'clock that Wednesday morning and started the crossing of the Atlas mountains in darkness. This was the leg he feared most; south of the mountains there was no pinpoint of any kind for hundreds of miles and he was glad that he would have daylight after all to help him find his way into Reggan.

One hundred and twenty miles to his right – if he was on track – lay Colomb Béchar and the Trans-Saharan motor track. As he flew south that motor track would come round in a great arc to meet him. It was still dark, and he had no cockpit lights, but he kept on steadily due south, lighting matches to check his compass direction every few minutes. Daylight came, and after five hours in which he had seen absolutely nothing to help him fix his position he finally spotted the white crease of the motor track converging on his own track from the right. There was no other road marked on the map and this must be the one he wanted. What he could not establish was his position along that road.

Directly ahead, as the road joined him and turned with him due south, he saw an airfield, and he decided to land there and find out where he was. He thought it was probably Adrar, and Adrar it turned out to be. This called for a change of plan. Adrar was only a hundred miles north of Reggan: by filling up with petrol here he could overfly Reggan and make straight for Gao, 800 miles distant and well within range.

He landed at Adrar at 8.30 that Wednesday morning and took off again 45 minutes later. He had completed what he had always felt to be the trickiest part of the flight. Ahead of him, winding away to his left now but never more than 40 or 50 miles to port of the direct track to Gao, lay the Trans-Saharan motor track. He could not afford to keep it in view the whole time, but he could always check his position by turning to port to meet it if need be, and if anything went wrong he would hope to have enough time to glide across towards it.

Unfortunately conditions were now very different from what they had been earlier that morning north of Adrar. A severe sandstorm was blowing up and visibility was reduced to almost nothing. In an effort to keep the motor track in sight, Lancaster turned too far to port, overflew the track in the sandstorm and continued on in a south-westerly direction until he picked up what he thought was the road again. In fact it was a more easterly road heading for Aoulef and In Salah on the trade route to the east. He landed at Aoulef to get his bearings at eleven o'clock and took off ten minutes later in the direction of Reggan, intending to pick up the main motor track again to Gao. But strong winds on which he had no possible check deceived him completely and the next thing he knew he was back at Adrar. This at least enabled him to pick up the right road again, but he followed it safely this time to Reggan. But the morning's flying had been disastrous. It was 3½ hours since he left Adrar and he had covered only 100 miles in that time The fuel he had used up in the process made it imperative that he land at Reggan, causing further delay. And the sandstorm was still blowing.

When he finally landed at Reggan he was worn out with frustration, nervously as well as physically exhausted. He was persuaded to rest while the plane was refuelled and while the sandstorm, perhaps, blew itself out, but he refused food. At four o'clock that afternoon, when he made ready to take off again, the wind was as strong as ever, and M. Borel, the head of the Trans-Saharan Company post at Reggan, strongly advised him to wait. It would be madness, he said, to attempt to get to Gao in these conditions, when even the beacons on the motor track would be impossible to see.

For an hour, two hours, Lancaster waited. His chance of beating Amy Johnson's record had almost gone. But he was still determined to press on and finish the course while the smallest hope remained. Then, as the wind dropped a little and the sandstorm eased, he decided to take off for Gao, still against the advice of M. Borel. It was 715 miles almost exactly due south and he hoped to get there about 2.30 next morning.

'We'll give you 24 hours,' said Borel. 'If we don't hear anything of you from Gao by this time tomorrow – say 6 o'clock tomorrow evening – we'll send a search party along the track. If you can burn something to light a beacon they ought to see you.' Borel gave him a box of matches, since Lancaster's own supply was running out, and a pocket torch so that he could see his instruments more easily.

At 6.30 that Wednesday evening Lancaster took off for Gao. He was still in a semi-exhausted state, and it seemed to the onlookers that he took off in an irregular, zig-zag fashion, as though he was almost too tired to control the machine properly. They watched him for some time after take-off, until the *Southern Cross Minor* became a speck in the failing light that finally disappeared, holding its course steadily due south.

Because of his recent navigational difficulties, Lancaster had made up his mind to follow the motor track for as long as he could see it. But within half an hour it was completely dark, absolutely black, and he could see nothing. For just over a hundred miles the direct course to Gao roughly coincided with the motor road. Then it struck across the open desert for 500 miles before picking up the road again 90 miles north of Gao. Perhaps after all that 500 miles of open desert presented the worst hazard, especially at night when he could not follow the road. If his engine failed anywhere on that vast stretch, his plane would be difficult to find. When one was talking in terms of thousands of miles, 500 sounded little enough. In fact it was an enormous stretch of brown, desolate, featureless sand, almost the length of the British Isles. To be forced down, even only 20 or 30 miles from the road, meant almost certain death.

The moon was not yet up and it would be pitch dark for many hours. Lancaster sat there in his tiny open cockpit a thousand feet above the invisible desert, flying his compass course, flashing his torch at intervals, trusting entirely to his instruments. He knew the road must be somewhere to his left, and he strained his eyes to try to make out the spindly line, but he could see nothing. Soon it would curve well away to the left, leaving him to plunge on across the desert. He had no check on his ground speed now, or his drift, so he was unaware that the wind was blowing from the north-east, hurrying him along in the right direction but tending to push him further and further away from the road.

An hour after leaving Reggan he was going well. He felt more relaxed now, ready to face the long hours of concentration that lay ahead. He was still drifting slightly to starboard but otherwise was making good his course for Gao. He would correct any inaccuracy when he got his pinpoint on the Niger.

At a quarter to eight that evening, Bill Lancaster began the long traverse of that 500 miles of open desert. He never completed it. Midnight came, then two o'clock, then three o'clock. Nothing was heard of him along the Niger, or at Gao. At last it was dawn, and although news was still expected of him at any moment, he was officially reported overdue. By midday it seemed clear that somewhere along that parched, inhospitable route Bill Lancaster had disappeared.

There was no great alarm at first. He could have come down to check his bearings, or to wait for daylight. He could be effecting some minor engine repair before completing the flight. He could have found succour at innumerable African villages north of the Niger. But the silence continued all day. Around the curve of the Niger from Gao to Timbuktu, and northwards from Gao to Reggan, every French military station was contacted by wireless. There was no news of Captain Lancaster. At six o'clock that evening M. Borel, as good as his word, sent the first car south from Reggan along the Trans-Saharan track to look for flares. The following morning the search was taken up from Gao.

The news that Bill Lancaster was missing travelled fast. It reached London later that day, Thursday, 13th April. But it aroused very little interest. No doubt it was only a scare; he would turn up in a day or so. These intrepid fliers made a habit of this sort of thing. There was much bigger news in the papers that Thursday evening, news of the Metropolitan-Vickers' engineers who were on trial for sabotage in Moscow; news of the tyranny of Adolf Hitler, who had just come to power in Germany; but above all, news of the Easter holiday, of race meetings and football matches, and of traffic jams on the roads. The forecast was for an Easter of brilliant sunshine, and everyone – or nearly everyone – was getting ready to enjoy it.

Lancaster's parents believed fervently that their son would soon be found. Chubbie Miller believed it too. But there were some who thought that if Bill Lancaster had really disappeared he had probably done it deliberately. Once he got behind Amy Johnson's record there had been nothing for him to live for. He had said so himself. There had been all that business of the murder trial. Very probably he had been guilty. This made it look very much like it. Perhaps in the circumstances it wasn't a bad way to go.

Lancaster had been flying for one hour forty minutes after leaving Reggan when his engine gave a warning cough. He was still flying his compass course for Gao, at an estimated 20 miles or so to starboard of the motor track. It was still pitch dark, and he flashed his torch round the cockpit to see that everything was in order and the fuel cocks were all turned on. Because of his heavy fuel load he had only climbed to 1,000 feet, but now he eased the stick back to gain height. Nothing further happened for about five minutes, and he was just beginning to relax when the engine coughed again and this time missed badly. At once he started losing altitude.

There was no time to think about gliding towards the road. He had to push the nose forward to maintain flying speed, feeling his way down, peering out of his open cockpit for a glimpse of the sand. Quickly he fastened his straps, in case he made a heavy landing. He also had the presence of mind to look at his watch, so that he could work out his position accurately once he was down. The time was 8.15.

Down and down he flew into the darkness, unaware how near the ground might be. When the machine finally struck he had still had no glimpse of it and he had had no chance to flatten out. The propeller splintered, the wheels of the fixed undercarriage thudded into the sand, and the aircraft bounced for 50 yards before falling again out of control and turning over on its back. Lancaster was hurled forward against his straps, ramming his head against the windshield, and the machine finally came to rest upside down. Lancaster was suspended in the cockpit unconscious, with gallons of petrol pouring from the ruptured tanks. Mercifully the plane did not catch fire.

Lancaster remained unconscious for several hours. When he came to he was still upside down in the cockpit and his first thought was to fight his way out. It was still dark, and his eyes were puffy and congealed with blood. He undid his straps and wriggled clear, scraping at the soft sand that had almost covered the cockpit. Eventually he managed to clear his eyes of most of the congealed blood and get them open.

His next thought was the drum of water. If that had been punctured like the fuel tanks he was finished. He crawled back to the cockpit and examined the drum. By great good fortune it was undamaged and the water was safe. He undid the fastenings and removed it from the cockpit. This at least would give him a few days of life.

The injuries and loss of blood had dried and caked his tongue and parched his throat, and he drank most of the

Thermos flask of water, nearly two pints, at one go. He was still feeling shaky, and his head ached, but he sat down on the wing of the plane, beneath what had once been the lower wing, and tried to think things out. With the coming of daylight he decided to make up his log. Keeping a diary had always been characteristic of him, and it had helped him to work out his problems. He would keep a diary now.

There was no chance of repairing the plane. If he could have made a reasonable landing he might have corrected the trouble – it must have been some fault in the petrol feed – and taken off again. But the *Southern Cross Minor* was in poor shape, and he wondered wryly what Kingsford-Smith would think of it now. Certainly to right the machine and effect repairs was altogether beyond the strength and capability of one man.

The French, of course, would look for him. They had promised to do so, and he was quite sure they would. But he had always known that if he was forced down on this leg any distance away from the road his chances of being found were remote. Perhaps it might be better after all to set out for the motor track.

He drew a diagram in his log-book of the motor track and estimated his own position in relation to it. He reckoned that he had been about 160 miles south of Reggan and 20 miles to starboard of the road when the engine cut. He had not been able to close the distance at all before the crash, so he had a 20-mile walk to reach the road. Even in the great heat that would be upon him when the sun rose it did not sound an impossible trek. He could manage one of the flasks, perhaps with difficulty both of them for a time, but he could not carry the 2-gallon tank. He doubted if he would be able to cover 20 miles with so little water, especially after his injuries. In any case it might well prove to be more than 20 miles, perhaps 25 or even 30. He had seen nothing since about half an hour after take-off and he could only make a rough guess at his position. He remembered his promise to Chubbie, and to his mother. He decided to stick to the ship.

He felt better for having made up his mind. Yet as the hours dragged by and the sun began to beat down out of a cloudless sky he began to realize that to stay put was virtually to resign himself to death. It gave him perhaps a week of life, if he rationed his water carefully, a week in which to burn flares at night in the hope that they might be seen from the road, and to watch for an air search by day. It would give him time to reflect, and the ability to leave a record of his last days. But it was very unlikely to bring rescue.

Inevitably his mind returned to the idea of the trek to the road. If he tried that and failed to intercept the car patrol he would be dead in a day or two. Then, if his plane was subsequently spotted from the air, he would have thrown his main chance away. Here he had water, and at least a little shelter, though even under the wing of the plane it was unbearably hot. His nose and eyebrows were badly cut and he needed to shield them as far as possible from the sun. In his present state he might never reach the road at all. Above him a lone vulture was circling, lower and lower, until he shouted at it and drove it away. No. His decision had been right. Stick to the certainty of a week of life. Stick to the ship.

They must realize now that he was missing. No doubt they would half expect him to turn up somewhere during the day, but by nightfall they would give him up. Where would they look for him, where would they imagine him to be? He wondered when the news would reach Chubbie, and his mother. Would Chubbie be able to get a plane to organize a search? He knew she would do all she could, and he had a warm feeling when he thought of her. His mother, too, he thought of with great affection: if only she could be reconciled to the fact that Chubbie was an essential part of his life.

The loss of blood had weakened him and he was half crazed with thirst already, yet he dare not drink more than a mouthful of liquid at a time. As the sun reached its zenith he became, without knowing it, slightly light-headed. He imagined that he might capture and tame the vulture, then leap astride it and fly with it to some green oasis. He saw it all with astonishing clarity, it looked so utterly real. Then he realized how fanciful it was. This was what faced him – a world of mirages and hallucinations, perhaps even of madness. He was afraid of blindness, too, because of the state of his eyes, and of infection because of his injuries, which were still congealed with blood and choked with sand. He could not spare any of his precious water to wash them clean, but he found a bottle of antiseptic in the first aid kit and emptied it over the cuts before bandaging his head, hoping to protect the cuts from sun and sand.

The hours of that first day dragged by interminably. The heat of the sun beat back from the sand like the glare of heat from an open furnace. He had taken all his external clothing off and he sat under the wing in his vest and shorts, still suffocatingly hot. All he had to look forward to was a week of physical torture culminating in a protracted and agonizing death. It would be the longest week of his life, as long as the rest of his life put together. In it, he knew, he would live his entire life over again. It would be a terrible atonement for any wrong he had done. Yet mentally he was neither depressed nor even unhappy. He adored Chubbie, and she returned his love. His parents had treated him wonderfully. Because of these things he knew a strange contentment. The love of these three people, father, mother and sweetheart, gave his life meaning just at a time when meaning might otherwise have been lacking. Whatever happened, he would not die unmourned.

He was buoyed up with the hope that Chubbie would try to do something to find him. When he thought about it he realized that alone she could do very little, but she would certainly do all she could to stimulate interest and inspire a search. Meanwhile there were the French motorized and air patrols. He must not allow himself to hope too strongly, but he must do all he could to help himself. He began to cut strips of fabric from the plane, rolling and wiring them into homemade flares. Then he doubled them over and hung them on the flying wires of the plane. After dark he would soak them in petrol one by one and burn them at intervals.

At half-past four that afternoon he saw a sparrow, and soon afterwards another one. He could not be so far from the motor track. At six o'clock the car would start from Reggan; but that was anything from 150 to 200 miles away. Even if it was the lower figure, the car could not pass him before half-past ten that night, probably much later, so it was pointless to light any flares before then.

His mind was clearer now, and he began to work out his water rationing plan in more detail. He had been taking sips of coffee at half-hourly intervals all day, and so far he had not touched the two-gallon drum. He decided to fill one Thermos flask a day with water from the tank. This way it would last for just over a week. It was no good trying to make it last any longer. In his semi-feverish state he might die of thirst if he didn't keep up a reasonable liquid intake, and any search that might be organized would give up after about a week.

At six o'clock he imagined the car starting from the outpost at Reggan. It would do well on that rough desert track to average 20 miles an hour. That meant another six, seven, eight hours. He looked again and again at his watch, but time seemed almost to stand still. The sun was going down and it was cooler, but the end of the day brought with it a profound depression. He had had practically nothing to eat since leaving Lympe two and a half days ago, and this contributed to his low spirits. He tried to eat one of the chicken sandwiches his mother had given him but it was rock-hard and he could not get it down. He couldn't even swallow the chicken. He badly needed some sort of stimulant to keep him going, and he tried a mixture of spirits of ammonia and half a pint of water as a pick-me-up. He felt better after this, and managed to find enough saliva to eat a small piece of chocolate. His faith in eventual rescue began to return.

It was no good giving up yet. He hadn't experienced a tenth of the misery he must endure if he was to hold out for as long as humanly possible.

The coming of night brought a marked change in temperature. He pulled on all his flying clothing but he still shivered with the cold. He had not expected to be so conscious of thirst when the sun went down, yet his craving seemed as great as ever. He sipped from his flask almost as frequently as he had during the day.

He kept his gaze almost unceasingly now on the direction of

the road, looking for lights, but he saw nothing. He examined his flares a hundred times. They were all ready, like a set-piece firework display. Promptly at half-past ten, just in case the car had made exceptional time, he lit the first flare. He was greatly heartened at its success. It burned for 60 seconds with a bright incandescent flame that must surely be visible for 20 or 30 miles. He went on burning them at regular intervals throughout the night, staring eastwards into the blackness for some sign that they had been seen. But when dawn came he was still sitting there alone under the wing, freezing cold and with a crushing sense of disappointment. There had been no answering signal from the direction of the road, and no sign at all of a search party. Yet some time in the night the car from Reggan must have passed him. They could not have seen him or they would have reached him by now. His flares must have been visible for many miles, so there was only one conclusion he could come to: he must be further away from the road than he had thought.

It looked as though his only chance of rescue was going to be an air search.

The French authorities, military and civil, were doing all that could have been expected of them and more. The Trans-Saharan Company's car left Reggan at the appointed time, but its occupants had so far seen nothing. An air search was being arranged south of Reggan. But the main area of search had been concentrated north of Gao and the Niger, at the terminal end of the flight. It was generally assumed that Captain Lancaster had probably completed the first part of the flight safely but been forced down in the later stages, either through fuel shortage or through failing to pick up the Niger. Relief planes were being drafted into Gao to join in the search.

In London, Chubbie Miller was finding it quite impossible to raise capital to organize a search in the time available. It was Good Friday, and London was deserted. The answer to all enquiries was that nothing could be done until after Easter. She could not even borrow a plane to fly out herself.

The Sahara was nearly 2,000 miles away. That meant –

except to a handful of trail-blazers – virtually the other side of the world. Very probably Bill Lancaster was already dead. That was the general attitude. Even if by chance he was still alive, any help that could be organized for him would come too late.

On that Good Friday morning, Bill Lancaster was taking stock of his position. He found that he had drunk a pint of water during the night, more than he had imagined, more than he could spare. He would have to make a similar amount last him through the day. Fortunately he had thought to refill his Thermos flask while the water in the tank was still ice-cold; that would make it more thirst-quenching and more refreshing.

He found some sun-burn preparation in the first-aid kit, and he treated his swollen eyes with it, painfully removing the bandage, which had stuck to his forehead. Otherwise he was in less pain than previously. He inspected his injuries in a metal mirror from the first-aid kit, and found that the cuts and bruises were quite severe but that so far they had not festered. No doubt these injuries would reduce his chances to some extent, but for the moment he still felt strong.

He had only a vague idea what his potential endurance ought to be, but he hoped to escape any serious weakness for at least another 48 hours. While he was still strong he would prepare as many flares as he could. He took the fabric he needed from what had once been the upper wing, now half-buried in sand. The lower wing, now uppermost, he left alone, fearing that if he tore the fabric from there it would make the machine harder to spot from the air.

He had been so optimistic last night, so sure that the car from Reggan would find him, that he had lit far too many flares. The box of matches was more than half empty: when he counted them there were only 18 left. He would have to light a fire tonight, and keep it going as long as he could, lighting his flares from there.

From half-past ten until half-past four in the afternoon the heat was a raging torrent. The comforting breeze of early morning dropped as the sun climbed the heavens. Water – that was all he could think of, all he could imagine. Every half-hour he took a sip from his flask, fighting with a conscious physical effort of will to prevent himself from tipping it up and swallowing a long, delicious draught. It was nectar and the thought of it dominated his whole being.

When the sun was at its hottest he broke the compass out of the instrument panel and tried the alcohol in it. It wasn't good, and he sprinkled the rest of it over his head, glorying in the cool sensation it gave to his scalp as it evaporated. But as the day passed without any sign of a searching plane his dejection grew. Why hadn't they seen him last night? Could he really have drifted so far from the road? A white butterfly, and then a dragonfly, tantalized him with their darting, fluttering message of an oasis near by.

Already he had prayed many times. He had gone down on his knees and thanked God for his deliverance after he had struggled out of the plane; it had been a miracle that he had not broken his neck, and that the plane had not caught fire. He had prayed for Chubbie and his mother. He had prayed for rescue, for the sight of a searching plane. Now he prayed for the strength to die like a man.

He was too exhausted that night to light many flares. Most of the night he slept. He tried to keep awake but couldn't. But he awoke on that Saturday morning refreshed and with renewed determination to conserve his energy, to last out as long as he could. He estimated that his water supply would hold out for another three or four days, provided he could keep a grip on himself and stick to his rationing plan. Every minute that he could hold out must increase his chances of rescue. They would know about his water supply, and they would expect him to last for a week.

That afternoon a wind laden with sand scorched across the desert, developing into a violent sandstorm. Lancaster wound his shirt round his head to protect his face, but the sand seeped between his parched, cracked lips and gritted against his teeth. His injuries were giving him acute pain, but there was no escape from the penetration of sun, sand and wind. He bore it as stoically as he could, and even in his diary he refused to complain at his plight. He had brought it on himself, there was no one else to blame, and he must play it out to the end.

Every half-hour throughout the day he took a sip of water. He drank two pints that day, and still had to fight to keep himself away from the tank and a third flaskful. When the sun went down he lit a fire from scraps of wreckage and pieces of propeller and burned several flares. He saw no answering signals. He had still eaten nothing since the crash, nothing in fact since leaving Oran nearly four days ago. He could not even swallow the chocolate now, but he melted some down in the metal Thermos cap, boiled some water, and made himself an excellent chocolate drink. Unfortunately it was too rich for him, or he drank it too quickly, and he could not keep it down. When the remainder – more than half – got cold he tried it again and this time managed to retain it.

During the night he had the most tantalizing experience of all. It began to rain, large raindrops as big as a penny and as cold as ice that fell with a plop on the sand-encrusted fabric of the wing. But before he could collect a single drop to supplement his own supply, the shower stopped as suddenly as it came, leaving nothing but a few tiny blobs of wet sand.

On the Sunday morning. Easter Sunday, he began to see signs of emaciation in his body; the skin was strangely shrivelled around his stomach and ribs. But with the deterioration in his body came a crystal clarity of mind. For the first time he remembered the details of his last flight clearly – the flight from Reggan. He wrote a full account of it in his log for the record. He recalled following the road until it got dark and then striking off across the desert. Somehow he must have drifted to starboard of his planned course. He saw the problem now from the point of view of the rescuers, the people who would be searching for him. They would not expect him to be far off course for the first few hundred miles. Further south they might search for him over a wide area, but 150 miles from Reggan they would not expect to find him far from the road.

He wrote about his situation with remarkable insight and lucidity, as well as with restraint and philosophy. He had his regrets, but no one was to grieve for him. Life in any case was a very short span in the wider scheme of things. He had not given up all hope, but his main concern now was his diary, and the last message he wanted it to contain.

The heat of the sun was appalling and his cuts worried him incessantly. He lay under the aeroplane wing, still sipping water at half-hourly intervals. The water-tank was now about half-full, sufficient for perhaps another three days. His mind returned again and again to the idea of the walk to the road. It was probably beyond him now, but he felt that he might have tried it that night but for his promises to his mother and Chubbie. It had been his last promise to both of them, and he must keep it.

Dusk was shading into night on that Easter Sunday when he saw it – a long stream of light to the east, falling lazily from the sky. He recognized it at once as a Very light, undoubtedly fired by an aeroplane. He could not see the plane, but he could still see the phosphorescent trail of the flare. It meant that the pilot would certainly see one of his if he fired it at once. He had a single remaining flare prepared, and he soaked it in petrol and lit it, watching it pierce the fresh darkness with a blinding flame that must be visible for many miles. He had no doubt that it must have been seen, and his fever-wracked body trembled with excitement. They had found him. In the morning they would come for him. It was a second miracle, and it was only just in time.

To celebrate what seemed his certain rescue, and to maintain his strength, he drank an extra pint of water above his daily ration. He could afford it now. He mustn't let his resistance get too low now that rescue was near.

He slept well that night, but awoke with a feeling of foreboding. Nothing had changed so far. Could the light have been a hallucination? Could he have dreamt it? In the early hours of Monday morning he suffered torments of doubt. He was positive he had seen the light. Whoever fired it must have been looking for him and must surely have seen the answering signal. Yet there was nothing to support his memory, no sign of a search party, no circling aeroplane. He regretted his impulsiveness in drinking that extra flask of water. It had cut his chances by almost a day.

His belief in rescue evaporated almost as quickly as it had been born. If that light had been what he had taken it to be he would certainly have seen an aeroplane this morning. Soon after ten o'clock, when the sun settled down to its ghastly overhead patrol, he had resigned himself again to his fate.

All day he lay gasping with thirst, but the memory of that extra flask drunk in the night still rankled and somehow he drank no more than his bare ration. At the end of that day he felt weaker, but he still had five pints of water, and he felt confident of lasting for two more days. As the light faded he scanned the eastern horizon and prayed that another Very light might appear in the sky. He had torn almost all the fabric off the Avro Avian, which now looked a dilapidated, moth-eaten wreck, hardly recognizable as an aeroplane, but he had several flares ready to fire, and half a dozen precious matches.

Darkness came and it seemed colder than ever. He dozed a little in the night, and woke up shivering. Rescue seemed unbelievable now, yet he still prayed for the miracle that would be needed to bring it about. Ahead of him lay his sixth day alone in the desert, and he braced himself to face the six hours when the sun was at its worst. He believed he could last out the day, but he was not sure about the morrow. He was troubled incessantly by flies, which worried away at the sores on his head and drove him almost to distraction. He was trying hard to remain rational, to keep his diary conscientiously, and to remember all the things he wanted to put in it. He had already given instructions about the diary itself; it was to go to Chubbie, and nothing in it was to be suppressed. He had put a note in it, too, for his estranged wife Katie and their two children, asking his mother to try and explain to them what had happened to his life, what had motivated him, the things he had held in his heart.

He was troubled by the knowledge that at this moment, and throughout the last week, he must be causing great mental suffering to his parents and to Chubbie. In his diary he asked their forgiveness.

The day passed, and night fell, and he hardly had the strength now to scan the horizon. All he could do for most of the time was to lie still and listen for the sound of aircraft engines that never came. Yet next morning he was still strong enough to write up his log in a firm and legible hand, steadier than he had been immediately after the crash. He had only two pints of water left, so this would be his last day.

He was conscious that he was writing what would probably be his last lucid entry in the diary. He must get all his message completed before he was exposed to the full heat of the sun. To Chubbie he spoke proudly of lasting a week in the middle of the Sahara with a crashed plane and a can of water. He would have done even better, he thought, but for his injuries. He was proud, too, of having kept his promise to stay with the ship. He urged her to give up flying, and to write her book one day, and dedicate it to him.

In a message to his mother he reproached himself for neglecting her for much of his life. He begged her to make her peace with Chubbie, to forgive and forget. He urged his parents to find solace in each other. His failure was no one's fault, just the luck of the game. He sent an affectionate message to his brother and a handshake to his father. He at least, he felt, would understand.

The sun was coming up, but he had one more thing to take care of before he crawled for the last time under the wing – his log book, the diary of his last days, whose preservation, now that his own hopes of life were gone, had become of intense importance to him. He was desperately anxious that his last messages should be delivered, and that the way he had faced his ordeal should be known to those who loved him.

It was not that he sought immortality through his diary. Many times in that lonely week he had asked himself the reason why, and he had arrived at a solution which satisfied him. To be loved was to exist. Therefore his existence would continue.

'The chin is right up to the last,' he wrote. 'I am now tying this log book up in fabric.' He wrapped it carefully in one of the last undamaged pieces of aeroplane fabric and secured it with a length of stay-wire before tying it firmly to the wing.

It was almost the last word from Bill Lancaster, but not quite. He wrote two more messages on the red, leather-bound fuel-card that he carried in his pocket.

'To my darling mother and my darling Chubbie,' he wrote. 'This is written on the seventh day away from Reggan. I hope you get the log and both read it together *for my sake*.

'No one is to blame. The engine missed. I landed upside down in the pitch dark, and there you are.

'See Nina and Pat for me. Kiss them. K will understand. I am tying the main log, which has all my thoughts and wishes expressed in it, to the strip of the left wing. I hope they find it and give it to you.

'Good-bye father, old man; write Jack, and good-bye my darlings. Bill.'

All that seventh day he lay under the wing sheltering under the last remaining strip of fabric. Somehow he survived the sun's merciless heat for yet another day, fortified at intervals by the last sips of water, until, by sundown, he had drained the last flask to the dregs.

Once again the night was bitterly cold. He did not expect to see the sun again. During the night he was delirious and his mind wandered. Yet when dawn came he still lay there under the wing, conscious and with a clear brain. Through all the interminable hours of hunger and thirst, heat and cold, hope and despair, he had had no reproaches for anyone, no self-recriminations. In his long ordeal of suffering he had found, in the final days, a tranquil and contented sense of destiny. He picked up his pen for the last time and wrote.

'So the beginning of the eighth day has dawned. It is still cool. I have no water.

'No wind. I am waiting patiently. Come soon, please. Fever wracked me last night. Bill.'

At half-past ten the sun climbed high into the heavens like a conqueror. But the victory had already been won.

The search for Bill Lancaster continued. Even on that eighth morning, when all the plans for a systematic search that had been instituted a week earlier had failed, General Georges, commander of the military forces in Algeria, called a further conference at his headquarters to work out a fresh plan in which planes and military stations hundreds of miles off Lancaster's intended route would co-operate. But they never found Bill Lancaster, or his plane. His complete disappearance remained a mystery.

Speculation about his fate continued. It was even rumoured at one time that he had been captured and was being held by Bedouin tribes. But the most popular theory, which was still being repeated many years later, was that Bill Lancaster had deliberately flown to his death.

The truth about the last week of Lancaster's life did not emerge for 29 years. Then, in February 1962, a French Army motorized patrol, pushing south from the desert atomic station at Reggan, found him and his plane in the heart of the Tanezrouft region of the Sahara, known to the Bedouin tribes as the 'land of thirst'. The skeleton of the aeroplane was still easily recognizable, and Lancaster's body, partly sand-covered, had been mummified by the extremes of heat and cold and was like a marble statue, so well preserved that the scars on his forehead from the crash were clearly visible. His right hand, frozen into 29 years of immobility when he was overcome by thirst, was still clutching at his throat.

Lancaster had been 170 miles south of Reggan and 40 miles from the motor track. He had come down at the one point on his 6,300-mile course where it might have been better for him to leave his plane and make for the road. He was tucked away in an area where no one had thought of looking for him. Searching planes had circled up to 60 miles either side of the road further south, but no one expected him to be so far from the road so early in his flight. Whether he could have reached the road, and whether he would have been found if he had, are questions to which there can be no answer.

The finding of the diary, perfectly preserved, cleared up not one mystery but two. It told the full story of Lancaster's last flight, but it did more than that. It told the full story of Lancaster. This story, like the diary referred to in the court at Miami, gave a glimpse into Lancaster's soul. From start to finish there is no mention in the log book of the murder trial. Yet everything else in Lancaster's life was there. Lancaster was not a man who would have suppressed his innermost feelings if he had retained any sense of guilt. All the indications from his past are that he would have poured out his heart in the diary. But he made not the smallest reference to the murder trial. He didn't even notice that, on the morning when he recovered consciousness upside down in the plane, it was twelve months to the day since he returned from St Louis to the bungalow in Miami.

The finding of Lancaster's diary, with its complete absence of any hint of obsession with the death of Haden Clarke, is the best evidence of all of the state of Lancaster's conscience, and of his innocence of murder.

BIBLIOGRAPHY

In addition to the sources listed in the Acknowledgements, the following books and magazines have been useful to me:

Winged Diplomat, the Life Story of Air Commodore Freddie West, VC, by P. R. Reid (Chatto and Windus).
The Return of Captain Hinchliffe, by Emilie Hinchliffe (Psychic Press).
Last Flight, by Amelia Earhart (Harrap).
Soaring Wings, a biography of Amelia Earhart, by G. P. Putnam (Harrap).
Daughter of the Sky, by Paul L. Briand, Jun. (Duell, Sloan and Pearce).
H.R.H. Princess Marina, Duchess of Kent, by J. Wentworth Day (Hale).
Looking Back, by the Duke of Sutherland (Odhams).
The Lady Be Good, by Dennis E. McClendon (John Day).
Flight 777, by Ian Colvin (Evans).
Sound and Fury, by Maurice Gorham (Percival Marshall).
All Weather Mac, by Captain R. H. McIntosh (Macdonald).
A Million Miles in the Air, by Captain Gordon P. Olley (Hodder and Stoughton).
My Fifty Years in Flying, by Harry Harper (*Daily Mail*).
Danger in the Air, by Oliver Stewart (Routledge and Kegan Paul).
Great Flights, by E. Colston Shepherd (A. and C. Black).
Aerial Wonders of Our Time, edited by Sir John Hammerton (Amalgamated Press).
The Second World War, by Winston S. Churchill (Cassell).
Flight.
The Aeroplane.
Life.
Reader's Digest.
Time.
Newsweek.
RAF Cranwell Magazine.

R.B.

Relive those dangerous days with
THE PAN BATTLE OF BRITAIN SERIES

Leonard Mosley
BATTLE OF BRITAIN (colour illus.) 25p
Richard Collier
EAGLE DAY (illus.) 30p
Richard Hillary
THE LAST ENEMY 20p
Elleston Trevor
SQUADRON AIRBORNE 25p
Air Vice-Marshall J. E. 'Johnnie' Johnson
FULL CIRCLE (illus.) 30p
Richard Townshend Bickers
GINGER LACEY: Fighter Pilot (illus.) 25p
Paul Richey
FIGHTER PILOT (illus.) 20p

Obtainable from all booksellers and newsagents. If you have any difficulty, please send purchase price plus 5p postage to P.O. Box 11, Falmouth, Cornwall. While every effort is made to keep prices low, it is sometimes neccssary to increase prices at short notice. PAN Books reserve the right to show new retail prices on covers which may differ from the text or elsewhere.